KIKKAWA, Jiro and Malcolm J. Thorne.   The behaviour of animals.
    Taplinger, 1972 (c1971).   223p   il   tab   bibl   70-172983.   7.95.
    ISBN 0-8008-0715-4
Kikkawa and Thorne are highly qualified to write a survey-type book
on the behavior of animals, starting with the most simple forms and
ending with the most complex. Their book, which is supposed to be
an introduction to the scientific study of animal behavior, is much
more than that since the authors place emphasis upon treating basic
concepts, as well as pertinent current theories.   The volume is illustrated with over 100 excellent illustrations that emphasize the topics
discussed. This well-written book will serve scientists, laymen, and
students in obtaining a grasp of that very complex attribute of life —
behavior. Recommended without reservation as a basic text on behavior and as a reference book for graduate and undergraduate
students.

The Behaviour of Animals

The Behaviour of Animals

Jiro Kikkawa D.Sc. *University of Queensland*
Malcolm J. Thorne Ph.D. *University of Queensland*

Taplinger Publishing·Company
NEW YORK

First published in the United States in 1972 by
TAPLINGER PUBLISHING CO., INC.
New York, New York

Published simultaneously in the Dominion of Canada by
Burns & MacEachern Ltd., Ontario

Library of Congress Catalog Card Number: 70-172983

ISBN 0-8008-0715-4

Front cover: White egret *Egretta alba* photographed
by Peter Slater

Contents

Preface

The study of animal behaviour is now becoming one of the major branches of life science. Among several reasons which explain this trend are:

- the study of whole organisms has produced a new framework of theory which allows investigations of exciting and intellectually challenging problems of social behaviour,

- the subject provides a valuable experience in integrating ideas of many different fields contributing to the study of behaviour, and

- the application of behaviour studies is proving essential not only in such fields as livestock management and social maladaptation, but also in recognizing the values of our environment and the changes in our philosophy and outlook towards life.

This book is an introduction to the scientific study of animal behaviour. It places primary emphasis on the attitude and methods of behaviour studies, while treating the basic concepts and some important current topics. It assumes very little previous knowledge of biology, no more than the treatment given in the most elementary science course at school, but makes no pretence of introducing an easy subject; understanding of animal behaviour in all its aspects is inherently difficult. At an elementary level we propose to examine the multidisciplinary origin of behavioural sciences, describe the control systems of organisms as they govern behaviour from the simplest to the most complex, and discuss mechanisms and functions of behaviour in terms of survival and successful reproduction. We feel that the comparative approach with emphasis on evolution—a classical approach in zoology—is still an important aspect of biology education to which students of behaviour should be exposed, particularly if they do not study other aspects of biology. We hope that the book will be useful to students of medicine, architecture, psychology and sociology, as well as biology students and those interested in animals in general.

We wish to acknowledge the aid of the following people who read parts of the manuscript and offered valuable comments: Maurice C. Bleakly, Douglas D. Dow, Don Morris and Carolyn Needham. During the initial stage of the preparation of this book, association with the National Science Curriculum Materials project proved valuable. The text also benefited from the responses of psychology and zoology students of the University of Queensland, who took our courses between 1965 and 1969. Any error or failure to communicate with readers is of course entirely due to us.

We are most grateful to Naoko Kikkawa who undertook the drawing of all diagrams. We also wish to thank Mr Crawford H. Greenewalt of Wilmington, Delaware, for providing the photographs of finches (Plate IV); Greg Gordon and Ross Robertson for making their unpublished work available for illustration (Figs. 5, 6 and 105); Sandra Thorne for indexing, and Odette Page-Hanify for typing notes and materials.

J.K. & M.J.T.

1 Introduction

The insatiable curiosity of children is often directed towards the behaviour of living things they see around them. The questions they ask about animal behaviour may not be answerable directly, but can often be reorganized to allow scientific inquiry relevant to the question. For example, the following questions are restated in a scientific manner and fall into different areas of study represented by more general questions.

What brings bees to flowers? Can butterflies discriminate between different kinds of flowers? How do ants locate food and transport it to the nest? Why do sheep congregate when they rest? How does a ewe tell her own lamb from others?

GENERAL QUESTIONS	AREA OF STUDY
● What makes an animal do what it does?	Study of causal relations
● How does an animal achieve what it does?	Study of mechanisms
● What purpose does such behaviour serve?	Study of functions
● How did such behaviour originate and develop?	Study of origins

Answers to these questions are the explanation of behaviour which may be found within the individual (*physiological*) or between individuals (*social*). The behaviour of animals is generally *adapted* for survival and successful reproduction of individuals, but as we shall see, this is not the explanation of behaviour.

A butterfly emerging from its chrysalis has no food in its gut. The absence of food in the crop is detected by internal sense cells and appropriate messages are sent to the brain. At the same time, the animal is receiving, via its external sense organs, information about the conditions in its immediate environment. If the temperature is low, flying may be impossible because the flight muscles must be warm to develop their full power. Co-ordination of the flight muscles is achieved in nerve centres in the thorax, but these centres themselves are influenced by messages from the brain and from

1

sense organs on the head. Thus we may say that the butterfly's behaviour is a series of *responses* to *internal* and *external stimuli*.

While looking for flowers (represented by the stimuli of colour and scent) the butterfly may be chased by a bird or a butterfly net. Its survival depends on the effectiveness of the evasive reactions to predators. It may also encounter another butterfly of the same kind and follow it. The response given to *other members of the same species* is thus different from the one given to *members of different species* (e.g. predators). When flowers of the right sort are found it will land on one, and by extending the long proboscis it will suck nectar from the flower. When the crop is filled with nutritious nectar its hunger (*a biological need*) will be satisfied.

In the course of this behaviour (feeding), the butterfly achieves, in relation to its environment, a number of things that are essential for the continuation of its life. Therefore, when behaviour of a whole organism is observed, we understand it as an expression of some co-ordination of physiological mechanisms within the organism. Such co-ordination of mechanisms is occurring at all levels on which life is organized (e.g. cells, organs, whole organisms).

PHYSIOLOGY ———— ETHOLOGY ———— ECOLOGY

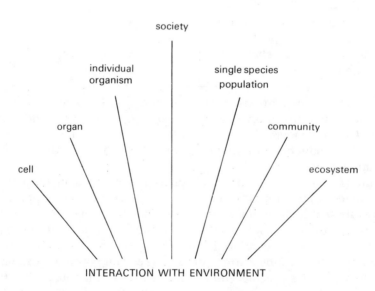

FIG. 1: Levels of biotic organization and behaviour studies

Individuals may form social groups, and populations of species may occur in nature, often showing an organization with many species (community). Responses of organisms to their environments are called 'behaviour' in its widest sense, and all living organisms 'behave' at molecular, cellular, organism, social and population levels to maintain their characteristic organizations in relation to respective environments (Fig. 1). Thus we may speak of, for example, behaviour of molecules in active cells or behaviour of rain forest under nuclear radiation.

Rudolf Carnap, an American philosopher of high attainments, suggested in 1938 in his *Logical Foundation of the Unity of Science* what he considered to be a new branch of science, embracing all aspects of behaviour at all levels of organization. He called it 'behaviouristics' to distinguish it from general biology. However, the studies of behaviour, its causal relations, functions, and evolutionary significance are restricted to the levels of whole organisms and social groups. This is the concept of *ethology* outlined as early as 1859 by a French naturalist, Geoffroy St Hilaire. At higher levels of organization, relations of organisms to their environments are studied under *ecology*, whereas those at lower levels are studied as *physiology*. Psychologists also study behaviour, but with the ultimate aim of understanding human behaviour. *Psychology* is naturally more concerned with mental processes and theories of learning.

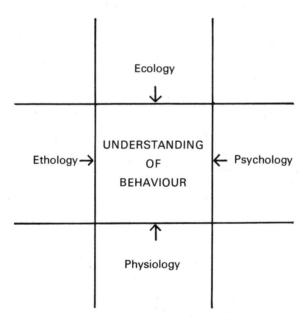

FIG. 2: Classical approaches to the study of behaviour

Today, the four fields that contribute to the study of behaviour, namely physiology (particularly neurophysiology), ecology, ethology, and psychology, have no distinct boundaries (Fig. 2). Each discipline has its own history of development and its own tradition of established methods. Since the object of *behavioural sciences* is to understand behaviour, there are naturally many possible approaches to achieve this object. Some approaches are of traditional kinds, while others develop as a result of interactions between different disciplines (*interdisciplinary* approaches).

2 Methods in Behavioural Sciences

EARLY HISTORY OF BEHAVIOUR STUDIES

EARLY NATURALISTS

As in other natural sciences, the histories of behavioural sciences demonstrate that observations alone do not form a science and that theories not based on careful observations are worthless.

The great modern tradition of natural history was founded by French naturalists about a century before the time of Charles Darwin. René Réaumur published between 1734 and 1742 six large volumes of *Mémoires pour Servir à l'Histoire des Insectes*, in which he gave detailed descriptions of the social life and parasitic, leaf-mining, gall-forming and other habits of insects. George Buffon studied habits of birds and mammals and published thirteen volumes of *Histoire Naturelle, Générale et Particulière* between 1749 and 1769. His ecological observations led him to believe that, given time, organisms under the influence of the climate change gradually. He influenced evolutionary ideas of Erasmus Darwin and Jean Lamarck.

In England, Gilbert White's *The Natural History and Antiquities of Selborne* . . . published in 1789 became a literary classic in this field. Following his *Origin of Species*, Charles Darwin published *The Descent of Man* (1871) and *The Expression of the Emotions in Man and Animals* (1873). In these he demonstrated that threat, courtship and other types of behaviour were products of natural selection, just as various organs of animals developed through natural selection.

European naturalists were still describing complex instinctive behaviour of insects (e.g. Fabre's painstaking observations) and giving novel accounts of animals found in remote areas (e.g. Brehm's *Tierleben*). However, when they accepted the concept of evolution they gradually applied the comparative approach of anatomists to the study of behaviour. They tried to demonstrate an evolutionary sequence of mental faculties and of animal societies. For example, Espinas (1878) discussed the evolution of animal societies. It seemed at last that behaviour observations meant more than mere description.

Unfortunately, however, many observations were not described in objective terms. Because expressions of animals sometimes remind us of our own emotions in similar circumstances, we may, for example, admire the seemingly courageous behaviour of the parent bird who tries to lure a predator away from the vulnerable young, or sympathize with the ant who drags a heavy load of food to the nest. This faulty attitude in science, called *anthropomorphism*, was commonplace in the description made by early naturalists. Darwin wrote, for example, that 'a pleasurable and excited state of mind, associated with affection, is exhibited by some dogs in a very peculiar manner; namely by grinning'. Although his detailed description of behaviour is still a useful reference, his anthropomorphic interpretation is not accepted today.

When animals see, hear or smell, the sensation they receive is different from the corresponding one of man. Similarly, when animals eat, drink or run, the motion used is different from that of man. Therefore, strictly speaking, the words that normally describe behaviour of man are not precise when used for the description of animal behaviour. Besides, senses and motions differ among animal groups, so that words applicable to one group may not be appropriate for another; for example, flies don't sniff and kangaroos don't gallop!

Similarity of particular behaviour between man and animals does not necessarily indicate similarity of underlying motivations and traits. Therefore, when anthropomorphic expressions are used today by animal behaviourists, they are used with the understanding that an interpretation is being made of the animal's motivations.

Early studies of animal behaviour also suffered from what are now called *teleological* explanations. Statements such as 'cockroaches seek shelter in order to avoid predators' or 'butterflies visit flowers because they want the nectar' are examples of teleological explanation which is simply a statement of presumed functions of behaviour and gives no consideration to causal relations. As we have seen, it is by the *response* of the butterfly to the stimulus of the flower that the butterfly is brought to the nectar. Therefore, 'because he wants the nectar' does *not* explain this behaviour. Many similar examples may be found in nature books for young children.

The fact that the response of the butterfly serves a useful purpose (getting food) means that the response is adaptive. Since the outcome of this response is not foreseen by the butterfly, it is called *directive* behaviour, in contrast to *purposive* behaviour in which the result is foreseen by the animal (e.g. a hungry dog coming to his feeding tray with anticipation of food).

Not all behaviour of man is purposive and not all adaptive behaviour of animals is directive. Since we cannot assume purposiveness in the behaviour of cockroaches or butterflies, teleology offers no explanation of their behaviour.

Biologists at the turn of the century condemned anthropomorphism and teleology, and described animal behaviour objectively. They sought mechanisms of behaviour in physiology, but avoided analysis of complex behaviour such as that found in the social organization of higher animals. The society as they knew it was based on unique characteristics of man, such as language, culture and history, so that the methods and concepts they developed were not considered applicable to animal societies. Any attempt at the demonstration of a biological basis of human social behaviour was condemned as *theomorphism*.

EARLY EVOLUTIONISTS

Following Darwin, many well-known biologists adopted evolutionary classifications of mental processes in animals and man. Among them were George Romanes who compared intelligence among animals, Herbert Spencer who developed a theory of animal ethics, and Lloyd Morgan who discussed mental evolution. They thought that instinctive acts could be modified by individual experience and that conscious and intelligent behaviour evolved from complex instinctive acts. These generalizations are now shown to be wrong. They did not know that stereotyped elements found in instinctive acts could not be modified by experience. Nor were they aware of the neural basis of behaviour that accommodated the evolution of intelligence. We now know that neither learning nor intelligence is likely to develop from complex instinctive acts.

MECHANISTIC PHYSIOLOGISTS

When an object approaches our eyes, our eyelids close. If we touch a hot or very pointed object we withdraw our hand rapidly. These *reflex* activities are simple, inherent and more or less invariable. They can, of course, be performed with different intensities, as you may recall from your responses to a slight pinprick and to a violent jab with a pin.

The internal pathway involved in reflexes consists of a sense organ (*receptor*) which sends its detected information via nerve cells (*neurons*) to the *central nervous system*. The incoming *sensory* neurons make functional contact with connecting neurons (*interneurons*) and these in turn connect with *motor* neurons. These leave the central nervous system and run out to the muscles or glands (*effectors*). In some cases, sensory neurons may connect directly with motor neurons making a short nervous pathway and allowing an animal to make a very rapid response to a stimulus (e.g. knee jerk in humans). The pathways just described are called *reflex arcs*.

Studies of reflexes showed that certain kinds of stimulus almost in-

7

variably evoked specific kinds of response. As a result, the reflex concept was expanded by physiologists to include instances in which the neuro-physiological evidence of nervous transmission (i.e. reflex arc) was lacking. This produced false unity of the observed behaviour. Plants grow away from gravity, frogs face light, and birds face winds when they land. Jacques Loeb studied such simple movements of organisms in relation to the external stimuli of light, electric current, gravity, temperature and pressure of the media.

In condemning anthropomorphism and teleology, Loeb put forward the theory that the responses of organisms are directed and, in fact, forced by these external stimuli. These responses were called *forced movements* (now known as *orientation movements*). He considered that these simple responses of both plants and animals were governed by the same mechanical laws, and therefore were not necessarily adaptive in function.

Complete physiological analysis of behaviour was not yet possible; even now it is limited to very simple responses involving only a few neurons. However, Loeb and other mechanists advocated that all instinctive acts could ultimately be explained as effects of reflexes and forced movements.

The mechanistic approach in the study of higher animals produced promising results in the analysis of learning processes. Ivan Pavlov, a Russian physiologist who won the 1904 Nobel Prize for his work on enzymes, experimented with dogs throughout his active life. He blew meat powder into the mouth of a dog and recorded the amount of saliva produced; then he sounded a bell each time he gave meat powder and repeated this procedure many times; eventually the dog salivated at the sound of the bell without the meat. In this case the meat was used as an *unconditioned stimulus* for salivation and the bell as a *conditioned stimulus* to produce the *conditioned reflex* of salivation.

Pavlov believed that the cortex of the brain played the essential role in this *conditioning*, and he promoted investigations to find the 'centres of learning' in the cortex. His work opened up a large field of study for psychologists who recognized the importance of learning in understanding human behaviour.

Through these studies, mechanistic physiologists fought against the *vitalistic* view of life, which considered the function of a whole organism to be greater than the sum of the functions of its parts.

BEHAVIOUR STUDIES IN LIFE SCIENCE

EXPERIMENTAL PSYCHOLOGY

Early comparative psychologists such as John Watson and Edward Thorndike were responsible for spreading the proposition that the process of learning is essentially the same in all animals. As a result, experimental psychologists elaborated apparatus to test learning abilities of one species — the rat. For example, Karl Lashley, who devoted his life to the study of the cortical functions of the brain, developed many techniques to test rats. In one of these, a rat was placed on a special jumping stand from which it had to jump through one of two windows. The windows carried different patterns which were to be discriminated by the rat. The rat had to 'make up its mind' before jumping, and the correct responses were rewarded with food on the other side of the window. Such a technique has been standardized for various tests of sensory capacities of animals and of the effects of brain destruction on learning abilities.

Experimental psychologists have concentrated their study on the laboratory rat; so much so that Frank Beach, a leading American psychologist, warned workers in 1949 that the Pied Piper's magic flute was now carried by the white rat, who played the music to lure experimental psychologists into extinction. Fortunately, comparative psychology has revived in recent years, and its field has expanded beyond theories of learning to include social behaviour and organization in different groups of animals.

SENSE PHYSIOLOGY

Jacob von Uexküll made an extensive analysis of animal senses and constructed the images of the world as perceived by different animals. The environment, in his view, can be defined only subjectively; there are as many different environments as there are different organisms. A bird in a tree sees a different world around it to the one seen by a caterpillar from the same position, just as the painter and the forester look at the tree from different points of view.

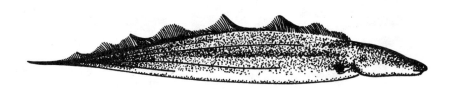

FIG. 3: An electric eel *Gymnarchus* showing electricity generating organs (lines on the side). The fish swims with the dorsal fin while keeping the body straight. Lissmann found that the constant electric discharge from the rear portion of the body creates an electric field with tail negative and head positive. Electric sensory organs are found on the head region. (FROM LISSMANN)

9

Donald Griffin was a university student at Harvard when he demonstrated that bats produced sounds not detected by the human ear. His experiments showed that bats avoided vertically strung wires in flight by using the ultrasonic echoes from the wires.

Echo-location by bats is one of many examples of incredible capacities of the sense organs possessed by some animals. An unusual method of finding objects is seen in certain fishes which possess an 'electric sense'. H. W. Lissmann, a Cambridge zoologist, found that the African electric eel *Gymnarchus* creates an electric field with the tail negative with respect to the head (Fig. 3). The conductivity of objects in the environment affects the distribution of the electric potential over the body surface, enabling the fish to detect the object.

In discrimination tests, animals are either rewarded with food when the response is correct, or punished, when incorrect, with devices which are painful or 'detestable' to them (e.g. mild electric shock). In such experiments an octopus was found to discriminate between the vertical and horizontal images of the same object. Bees, on the other hand, could recognize objects from the irregularity of their shapes regardless of position (Fig. 4).

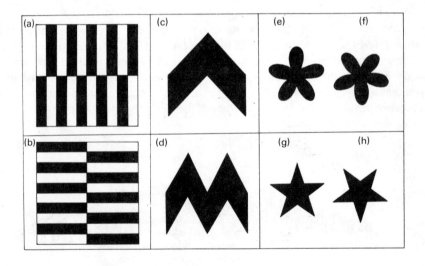

FIG. 4: Comparison of discrimination by the octopus and the bee: Sutherland found that octopuses could discriminate two similar figures if the orientation of pattern differs (a and b), but could hardly distinguish between different figures with apparently similar orientation (c and d). On the other hand, bees could discriminate two similar patterns (c and d, e and g, f and h), but not the same figures differently oriented (a and b, e and f, g and h).

NEUROPHYSIOLOGY

The understanding of behaviour within the individual organism came from the study of nervous mechanisms (*neurophysiology*). The history of neurophysiology shows a parallel development to the progress in electric and electronic technology. For example, galvanometers were used to show that nerves and muscles generate electromotive forces. When the oscilloscope and powerful amplifiers were invented, the electrical activity of cells could be measured for the first time.

Ever since the general anatomical connections between receptors and effectors were discovered, the ways in which messages are transmitted in the living body have been an important field of study. Use of micro-electrodes in experiments made it easier to record bioelectric signals, measure potential differences, and stimulate nerve and muscle cells. Through such experiments the nature of effective stimuli to different kinds of nerve and muscle fibres was revealed. When considering the mechanism of transfer of messages between nerve cells, it is important to know whether the contact between the cells allows the nerve impulse to pass without physical interruption from one to the other, or whether some other agency is involved in transferring excitation. When such speculations were first made, nobody had been able to actually observe the contact point under the microscope. The development of the electron microscope made it possible to examine 'contacting' cells, and the view that they were actually separated by minute gaps was confirmed. With the development of electronics, the electric and chemical transmissions across the nerve connections and the electronic transmission across the cell membrane began to throw light on the nature of messages transmitted within the living body.

The advancement of neuro-anatomy in recent years was also due to the development of electrical and chemical micro-techniques. These helped to map out regions of vertebrate brains associated with specific functions underlying behaviour.

However, the co-ordination of an animal's activities is not achieved entirely by nervous mechanisms. Chemical co-ordination is also responsible for the control of activities.

ENDOCRINOLOGY

The influence of hormones on reproductive and aggressive behaviour has been studied by injecting animals with glandular extracts or substances that stimulate the activity of particular glands. Sometimes particular glands were removed surgically to study behaviour of the animals lacking them.

Daniel Lehrman studied reproductive behaviour of the ring dove for many years. He found that caged ring doves show a regular cyclic activity

11

of courtship, nest-building, egg-laying, incubation, feeding of the young, and then courtship again. This cycle is normally repeated every six weeks, but if birds are kept singly the cycle does not occur. Lehrman induced nest-building behaviour and sexual receptivity in isolated females by injecting estrogenic hormone and progesterone. These hormones are normally produced by the gonads under the control of other hormones secreted by the pituitary gland. The pituitary gland is, in turn, influenced by the hypothalamus in the brain. Therefore, pituitary secretion can occur in response to external stimuli via brain mechanisms, and can influence behaviour by controlling secretion of hormones from other glands. Nest-building and copulation in birds generally coincide with the rapid growth of the ovarian follicles and the oviduct. Full oviduct growth requires the successive action of estrogen and progesterone secreted by the ovary. If the ovary is removed, the bird will not respond to the courtship of a male. Similarly, castrated males do not initiate courtship at the sight of a female.

In order to test the effect of the partner's behaviour on the egg-laying behaviour of females, Daniel Lehrman and Carl Erickson placed forty female doves in separate cages, each cage with a view of a separate male through a glass plate. Of the forty males used as stimulus animals, twenty were normal and exhibited vigorous bowing and cooing upon seeing a female, whereas the remaining twenty had been castrated and did not exhibit courtship behaviour. During the seven days following the introduction of males, ovulation occurred in thirteen of the twenty females with normal males, and in only two of those with the castrates. Development of the ovary is not induced merely by seeing another bird, but by seeing and hearing the male in courtship. Similarly, incubation and feeding of the young are responses to powerful stimuli from the eggs and the young respectively. Thus, synchronization of cyclic behaviour between the paired ring doves is due to the effect of the partner's courtship and other external stimuli acting through the brain and endocrine glands.

As in neurophysiology, the sort of question asked by early endocrinologists was 'Which stimulus (hormone in this case) produced which action?' The answer was not simple, because complex feedback systems are involved in behaviour mechanisms, producing much variation in the response patterns of animals. As shown in the ring dove experiments, external stimuli are often involved in the feedback systems acting through the central nervous system and modifying the effects of endocrine secretion. Hormones act on specific target cells within the living body, but the effect of any particular hormone on behaviour varies greatly. It depends on the nature and intensity of the secreted stimulus, and what other hormones are acting and interacting at the same time.

Co-ordination and integration of complex behaviour remain largely

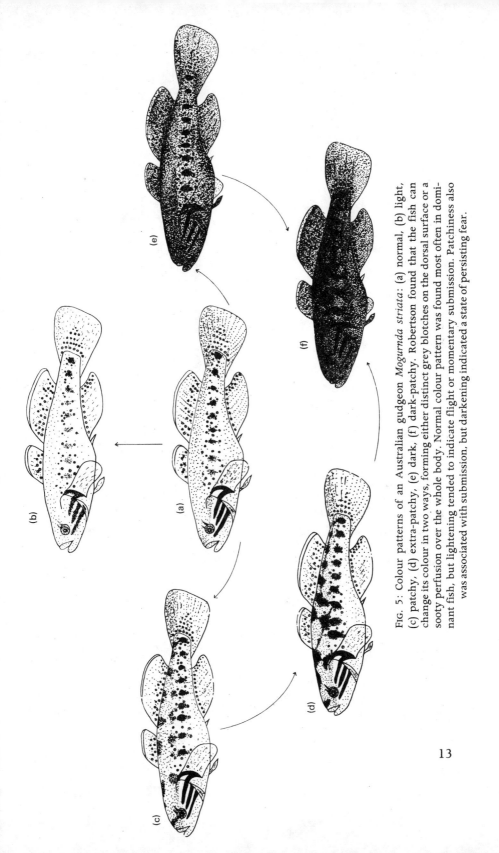

FIG. 5: Colour patterns of an Australian gudgeon *Mogurnda striata*: (a) normal, (b) light, (c) patchy, (d) extra-patchy, (e) dark, (f) dark-patchy. Robertson found that the fish can change its colour in two ways, forming either distinct grey blotches on the dorsal surface or a sooty perfusion over the whole body. Normal colour pattern was found most often in dominant fish, but lightening tended to indicate flight or momentary submission. Patchiness also was associated with submission, but darkening indicated a state of persisting fear.

13

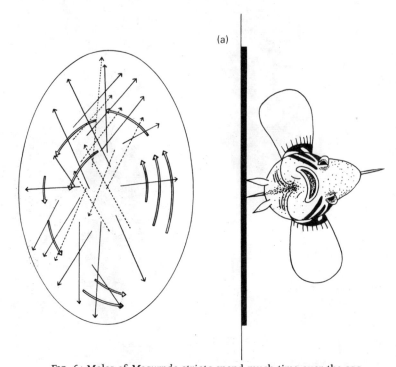

(a)

FIG. 6: Males of *Mogurnda striata* spend much time over the egg
mass, protecting it. They also move while keeping contact with
eggs, alternately beating pectoral fins (fanning). Fanning creates
a current, helps aeration and prevents silting of eggs.
(a) The method of fanning, with a fish covering eggs on a vertical
surface. The fish moves forward (single unbroken line), backward
(broken line), or sideways (double line) while fanning.

unexplained in physiological terms. Undoubtedly all behaviour patterns
have a physiological basis, but for many of them the only means of approach
available at present is the study of whole organisms.

ETHOLOGY

The *colour patterns* of the Australian gudgeon *Mogurnda striata* are
shown in Fig. 5. Ross Robertson studied the behaviour of these fish and
found that they can change their colour in two ways, forming either
distinct grey blotches on the dorsal surface or a sooty covering over the
whole body. Dominant fish have the normal colour pattern but fish behaving
submissively often develop grey blotches. Momentary submission or flight
tends to lighten the colour, but with sustained extreme submission the
colour darkens.

14

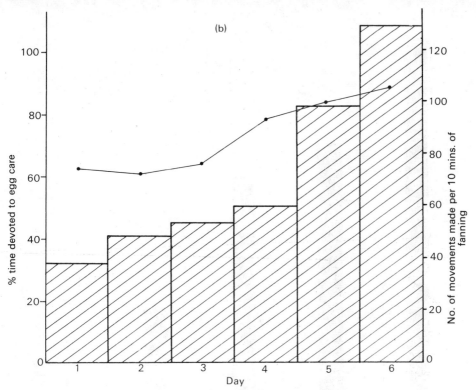

(b) Robertson found that the proportion of time a male spent for
egg care (line graph) and the number of movements he made
during fanning (shaded histogram) increased considerably from
spawning (day 1) to hatching (day 6)

Male gudgeons spend much time protecting the egg mass. They also move
about while keeping physical contact with eggs, alternately beating the
pectoral fins. This *fanning* movement of the fish is a *motor pattern*. It creates
a current, helps aeration, and prevents silting of eggs. The method of
fanning is illustrated in Fig. 6a with a fish covering eggs on a vertical surface.

Such objective descriptions of colour patterns and motor patterns are
classified by ethologists, usually in *functional contexts*—for example
maintenance behaviour, reproductive behaviour and aggressive behaviour.
In the above example, changes of colour are part of the maintenance
behaviour, whereas fanning is part of the reproductive behaviour.

Further objective description of behaviour can be made in quantitative
terms. Robertson measured the proportion of time a male spent caring for
eggs and the number of movements made during fanning. The results
presented in Fig. 6b show that both increased considerably from the day of
spawning to the day of hatching. Thus, quantitative descriptions can be
made by recording the number, duration, frequency and magnitude of any

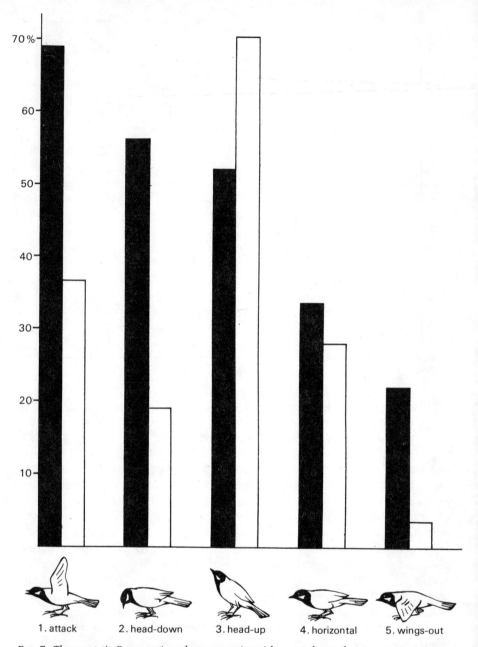

FIG. 7: The great tit *Parus major* when aggressive either attacks or threatens an opponent. Using five hand-reared birds, which showed aggressive responses to a paper roll and fleeing responses to a small electric lamp, Blurton Jones tested responses to 'attack stimuli' (paper rolls of various colours) and to 'attack plus fleeing stimuli' (paper rolls plus a lamp). He tested single birds for 10 seconds, recording presence or absence of attack (1) and threat postures (2–5). It is possible for one bird to assume several postures in this period, so that each frequency was independent of the others. Comparing frequencies (expressed in the columns as percentages of the total) obtained from 235 tests using 'attack stimuli' alone (black), with those obtained from 104 tests using 'attack plus fleeing stimuli' (white), he found that the two situations produced different response patterns. He had predicted that reduction in attack would be accompanied by increase in threat displays, but the results shown here were contrary to the prediction. (FROM BLURTON JONES)

given behaviour, and these may be analyzed in relation to internal and external stimuli.

Recently, Blurton Jones at Cambridge examined the relations between the various types of threat postures exhibited by five hand-reared great tits, *Parus major*. These birds showed aggressive responses to a paper roll and fleeing responses to a small electric lamp. He took advantage of this habit and used paper rolls of various colours as 'attack stimuli' and the lamp as a 'fleeing stimulus'. A combination of both types of stimulus was used in the test to produce 'attack plus fleeing stimuli'. Every bird was exposed to each stimulus for ten seconds, and he recorded the presence or absence of 'attack' and of the various 'threat postures' shown in Fig. 7. He had predicted that birds would show more threat postures if attack was suppressed by fleeing stimuli, but the results shown here were contrary to his prediction.

This is an example of the study of underlying motivations in the behaviour of birds. A given condition or stimulus may be reproduced artificially by controlling experimental situations or presenting various models to the animal. If the animal behaves as predicted, its motivation may be explained.

The results of another experiment shown in Fig. 8 illustrate the type of experimental design which allows us to postulate the function of social behaviour in terms of the survival of individuals. The birds used in this experiment were Australian silvereyes, *Zosterops lateralis*. Two groups of ten birds were kept in similar cages over one winter. One group was used as a control and was fed from two feeding trays. The other group (experimental) was fed from a single tray with a *similar amount* of the *same food*. The experimental group was deprived of food for one hour each morning and thus stimulated to fight over food after each deprivation period. The birds were weighed at the beginning and end of the experiment and the percentage weight gain of each bird was calculated. The results are plotted in Fig. 8 against the social rank of individual birds. The control group gained significantly more weight than the experimental group. This suggests that the stress due to starvation and the increased social interaction due to limited distribution of food affected the weight gain of certain birds in the experimental group. If a similar situation occurred in the field or in domestic animals, natural or artificial selection would eliminate individuals with certain behavioural characters with respect to aggression.

In the course of such analyses, ethologists identified many types of behaviour (*behaviour patterns*) which were specific to a particular species of animal (*species-specific*). These behaviour patterns are little influenced by experience and are therefore called *innate*.

Since the turn of the century the word 'instinct' has been shelved by

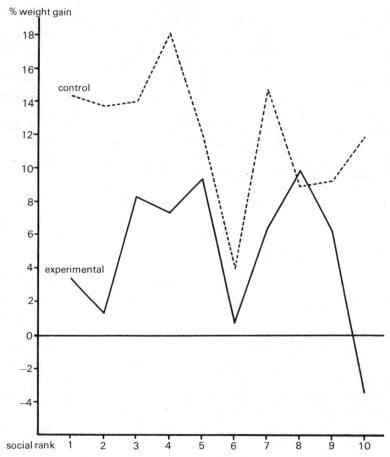

FIG. 8: The two groups of 10 silvereyes *Zosterops lateralis* were kept in similar cages over one winter. One group (control) was fed from two feeding trays, whereas the other (experimental) was fed from a single tray with a similar amount of the same food. The experimental group was deprived of food for one hour each morning and thus stimulated to fight over food after each deprivation period. Birds were weighed at the beginning and end of the experiment and the percentage weight gain of each bird was calculated. This is plotted in the figure against the social rank of the bird. The control group gained significantly more weight than the experimental group, suggesting that stress of food deprivation and increased social interaction due to limited distribution of food affected the weight gain of certain birds in the experimental group. (FROM KIKKAWA)

experimental zoologists because it was not definable in physiological terms. Similarly, psychologists did not recognize instinct as unmodifiable behaviour. Two European ethologists, Konrad Lorenz and Niko Tinbergen, revived the concept of instinct in the late 1930s. They developed the concept of *drive* in its place to explain behaviour associated with 'instincts', and considered that different hormones controlled different drives. Independent drives are related basically to nutrition, reproduction, maintenance (e.g. grooming and preening), sleep, social relations and possibly aggression. Different drives may be activated in an animal at different times of the day or year, sometimes with periodic changes. The state of activation of a particular drive is called *motivation*.

When a very hungry mouse starts feeding it will not easily change its activity. While its hunger drive is strong, ordinary external stimuli will not affect its activity. If a predator attacks, however, such a powerful external stimulus overrides hunger motivation and the mouse flees.

Ethologists considered that each independent drive had a specific outlet in behaviour (e.g. hunger drive—feeding). This *specificity* of drives was associated with distinct *action patterns*. Action patterns are often fixed according to the drive and according to the species. They are therefore referred to as the *fixed action patterns*, and are considered to be the core of instinctive behaviour. A mechanism to prevent the continuous discharge of drives is believed to exist in the central nervous system. If a hungry animal cannot find food and discharge its hunger drive, the tension in the central nervous system will build up. This build-up of tension will lower the threshold for the release of a particular action pattern, for example feeding, which belongs to the hunger drive. Such building-up of tension is related to an accumulation of hormones in the central nervous system. If this process continues too long, the drive may be discharged with very little or practically no external stimulus. For example, a ewe develops a strong maternal drive before giving birth to her lamb, and may adopt a newborn lamb nearby while her own is actually being born. Such an activity is called *vacuum* or *overflow activity*. In hoofed animals the female normally leaves the herd before giving birth, so that the parental drive is denied its discharge until her own young appears. A virgin bitch lavishing her maternal care on a bone, or thumb-sucking behaviour of infants, also illustrates overflow activity of instinctive behaviour. In these cases a substitute is used for the discharge of drives, and in forming a habit such substitutes become external stimuli. As a result a new, learned drive (*secondary drive*) may appear. There are instances, however, in which motor patterns appear without any external stimulus at all. For example, in the absence of flying insects, caged flycatchers often go through all the movements of catching and eating an imaginary fly.

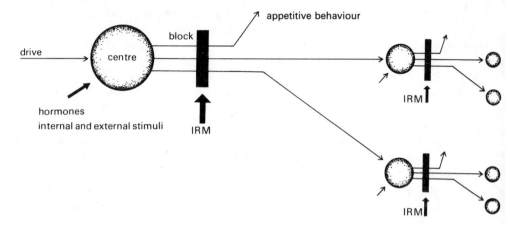

FIG. 9: Part of a hierarchical model of innate behaviour mechanisms (AFTER TINBERGEN): Tinbergen has postulated that at each centre a block prevents continuous discharge of the drive, and an innate releasing mechanism (IRM) when activated will remove the block

If the action patterns constitute an end-point in a chain of instinctive behaviour they become *consummatory acts*, which normally attain a final, biologically desirable goal. The stimulus that provides the appropriate situation for the motivated animal to discharge its drive is called the *sign stimulus*. Sign stimuli release fixed action patterns by triggering an *innate releasing mechanism* in the central nervous system (Fig. 9).

Ethologists have now discovered and analyzed many complex innate behaviour patterns in various groups of animals, particularly in insects, fishes and birds. Some of these will be discussed later.

BEHAVIOUR GENETICS

Selective breeding of sheep dogs, race horses and fighting bulls shows that innate behaviour has a genetic basis. Biologists, also, select rats, mice and the domestic fowl for certain behavioural traits such as aggressiveness, particular learning performances, and special capacities of sense organs. However, mechanisms of the inheritance of behaviour are not fully understood. The study of the genetic basis of behaviour is called *behaviour genetics*, a new and developing field. Behaviour geneticists analyze behaviour of hybrid animals obtained by cross-mating different inbred strains, different breeds, or different races found in the field.

The use of inbred strains makes genetic control possible, but at the same time limits the number of genotypes that can be handled. Besides, any particular genotype represented by a particular strain produced in the laboratory would never occur under natural conditions. Therefore, the generality of genotypic effects on behaviour or the degree of genetic determination is very limited. In special cases, where single-gene mutations affect both morphological and behavioural traits, it is possible to observe genotypic effects on behaviour. For example, in mutant forms of *Drosophila melanogaster*, the recessive gene that produces white eyes reduces the frequency of copulation in homozygous males by about 25 per cent, thus decreasing the frequency of this gene in subsequent generations.

Psychologists have demonstrated pure genetic control of many behaviour traits in rodents—for example, variations in normal sexual behaviour in guinea-pigs, and alcohol preferences in mice. These and other genetically controlled behavioural traits show quantitative differences between different strains of animals, and in some cases their associations with physiological or biochemical characteristics of the strains have been demonstrated. However, precise genetic analysis of rodent behaviour is much more difficult than that of *Drosophila* behaviour.

Modification of behaviour through experience is an adaptive response of animals and is inevitable even in a controlled experimental population, making genetic analysis of behaviour extremely difficult. For most behavioural characters observed in the field populations there is, as yet, no means of analyzing their genetic basis.

SYSTEMATICS AND EVOLUTION

As we have seen, some behaviour is innate and species-specific. Such behaviour provides information on the evolutionary relations of animals and can be used to supplement studies of morphology.

For example, birds scratch their heads in two stereotyped ways. The first is to lower one wing and bring the corresponding leg to the head over the shoulder (scratching over the wing). The second is to bring the leg straight up in front of the body (scratching under the wing). As a rule, no species employs both methods, but within taxonomic groups the method used is uniform. Pigeons, gulls and quails all scratch under the wing, whereas nearly all perching birds (passerines) scratch over the wing. Among waders, all plovers scratch over the wing and all sandpipers scratch under the wing. Among parrots, lorikeets and cockatoos scratch 'under' while grass parrots and rosellas scratch 'over'. Therefore, the method of head-scratching is a behavioural character shared by all species of certain groups. Since the classification of parrots is not easy using morphological characters, such behavioural characters provide useful information for the study of evolution.

In the birds of paradise and bower-birds, the types of display are actually used in the groupings of genera and subfamilies. Among closely related species, differences in the details of courtship behaviour act as important mechanisms of reproductive isolation. This is particularly important if the species occur together in the same general area. This has been demonstrated in many insects, freshwater fish, frogs and birds. Some of these species could not be distinguished on their structure alone, but differ in mating calls (e.g. crickets, frogs), nesting behaviour (e.g. cave swiftlets), or annual cycles of behaviour in general.

While observing the Australian raven in the field, Ian Rowley noticed that two distinct groups of ravens in one area showed different types of behaviour. He thought there might be some morphological differences between the two groups, but he could not distinguish the two among museum specimens. Further studies of breeding behaviour showed that they belonged to two distinct species (see table).

BEHAVIOURAL DIFFERENCES BETWEEN TWO SPECIES OF RAVEN

	Corvus coronoides	Corvus mellori
Territory size	300 acres	10 acres
Activity inside territory	all activities	nesting
Territorial advertisement	distance signal: long drawn-out call from conspicuous perch in territory	short-range signal: sharp call from any-where, with distinctive wing flip
Foraging	within territory	outside territory
Courtship	aerial chase	ground promenade
Nest-site	position in tree more than 13 m above ground, commanding good all-round view	position in tree less than 13 m above ground or on ground, often hidden in thick cover
Commencement of breeding	mid-July	early August
Period of young in nest	45 days	36 days
Young leaving nest	remain in territory for at least 4 months	move, with parents, to flock after 2 or 3 days
Movements	adults—resident; immatures—nomadic flocks	all birds in nomadic flocks in non-breeding season

A comparative approach in the study of behaviour revealed some possible courses of evolution taken by selected species groups. These groups contained a behaviour series from simple to complex, or generalized to specialized. For example, the behaviour of ants or parasitic wasps can be arranged according to increasing complexity. Such arrangements are generally considered to reflect the general trend of evolution. The different types of bowers built by bower-birds are shown in Plate I. These can be arranged from the simple to the complex types: the 'arena' type, such as one built by the tooth-billed bower-bird, may be considered primitive, while the 'avenue' type, such as one built by the spotted bower-bird or the golden bower-bird, may be considered to have evolved from the 'mat' type. However, there is no proof that the 'mat' type did not evolve independently. Because behaviour leaves no fossils, there is always a danger in interpreting the evolutionary sequence of behaviour in present-day animals.

ECOLOGY

Population ecology is concerned with the study of the regulation of animal numbers. Population density changes according to the rates of birth, death and migration, and population ecology deals with the factors that affect these rates. Many ecologists have been engaged in the study of behaviour which varies with population density and contributes to changes in the above rates.

Migratory locusts and lemmings become extremely active as population density increases. Flour beetles become cannibals if kept in crowded conditions. Male rabbits or rats born into very dense populations are more aggressive and more active sexually than those born into less dense populations. As these and other social behaviour patterns tend to act as a limit to the population size, some ecologists believe that these behavioural attributes have evolved to regulate the population size. Other types of behaviour that appear in response to certain physical factors of the environment are more obviously adaptive. These include activities having periodic cycles associated with light, temperature or tide on the one hand, and cycles of reproduction or migration on the other. These activities are subjected to analysis by ecologists. They also study behaviour associated with the selection of suitable habitats and sites which provide headquarters for such activities as sheltering, roosting, nesting, display and feeding.

In Australia, many species of animals are adapted to arid areas where suitable conditions for breeding occur only sporadically. Red kangaroos, wood-swallows, budgerigars and many water birds show nomadic movements, whereas some frogs burrow down to 150 cm in soil in the dry season.

Non-breeding members of Australian magpies are often pushed out of the breeding territory, whereas those of the white-winged chough and the kookaburra participate in the rearing of young. In blue-wrens the young birds from the first clutch may help parents in collecting food and feeding the young of the second clutch. In good conditions young zebra finches from the first clutch mature rapidly and produce their own progeny in the same breeding season. These social systems are behavioural adaptations for adjusting the breeding capacity of the population to the capacity of the environment.

3 Control Systems of Animals

For an animal to function efficiently, its activities must be controlled and co-ordinated. There are two main types of control systems found in animals: the *nervous system* and the *endocrine system*. Neither system occurs, however, in two phyla (the protozoans and sponges).

These two systems may superficially appear to be separate, but the boundary between them is not clear. Both systems rely on the production of certain chemicals (sometimes identical ones) which function as 'information carriers', but with the nervous system there is the added incorporation of electrical signals to convey information from one part of the body to another. Many parts of the endocrine system probably evolved from nervous or pre-nervous tissues, and in many animals the majority of their endocrine glands are simply accumulations of nerve cells which are modified for secretion.

PRIMITIVE SYSTEMS

Unfortunately, the co-ordinating mechanisms of protozoans and sponges are not very well known, and what little is known does not contribute markedly to our knowledge of the evolution, development and physiology of nervous or endocrine systems. Since these two phyla lack nervous and endocrine systems, the possible relation between their co-ordination mechanisms and those of the more highly organized animals may be questioned.

PROTOZOA

These microscopic animals are acellular (or can be considered as unicellular) and occur in moist habitats (fresh and salt waters, damp soil, inside other animals, etc.). There are many thousands of different species of protozoans, which have been conveniently divided into groups on the basis of their method of locomotion. Some put out finger-like processes with the rest of the body flowing behind (*rhizopods*), some have one or more long, whip-like, protoplasmic extensions by which they swim (*flagellates*) or have the body covered with numerous short processes used for swimming (*ciliates*), while others lack any such processes in the adult phase and are parasitic in habit (*sporozoans*).

25

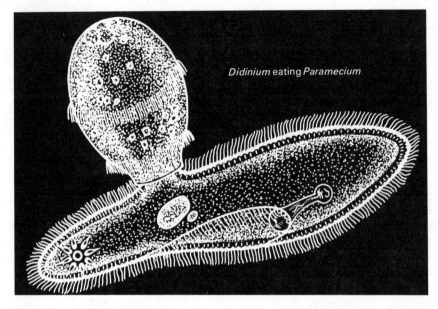

Didinium eating *Paramecium*

In the very simplest forms, such as *Amoeba*, adaptive responses to a variety of stimuli are shown. *Amoeba* will move away from a brightly lit area, a concentration of injurious chemicals, or mechanical prodding. It will move towards and engulf food material by flowing around all sides of the object. But, in essence, these responses are either *approach* (to food and some chemicals) or *withdrawal* (from adverse stimuli), and *Amoeba* appears incapable of more complex types of behaviour.

The ciliates often show rapid adaptive movements or direction changes to a number of different stimuli. Predatory forms, such as *Didinium*, respond in different ways towards potential prey species, indifferent species and adverse stimuli.

Paramecium shows a well-marked *avoiding reaction* to extremes of temperature, injurious chemicals, objects in its path, etc. The animal stops moving, reverses the normal beat of its cilia, and as a result moves away from the stimulus. A slight turn is effected and the animal moves forward in its new direction. The same reaction is repeated until the animal no longer encounters the adverse stimulus. Such a reaction involves the co-ordination

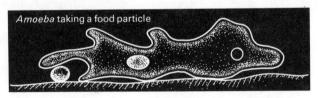

Amoeba taking a food particle

of thousands of cilia clothing the exterior of the animal. Yutaka Naitoh and Roger Eckert recently found that if the anterior part of *Paramecium* is stimulated mechanically, the permeability of the cell membrane to calcium ions increases. The cilia reverse their beat due to the resulting changes in the ion balance and electrical charges across the cell membrane. Stimulation of the posterior part of the cell increases potassium permeability and the cilia beat more rapidly. Thus an avoiding reaction is shown if the anterior end is stimulated, while the animal moves fast in a forward direction if the posterior end is stimulated, as shown in Fig. 10.

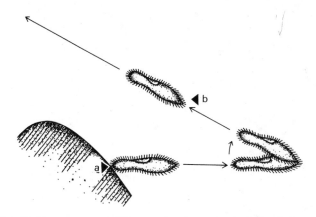

FIG. 10: Two responses of *Paramecium* to mechanical stimulation:
(a) mechanical stimulation of the anterior end → increased Ca^{++} conductance of cell membrane → depolarization of membrane → electrotonic spread of depolarization → reversal of ciliary beat → reversed direction of locomotion
(b) mechanical stimulation of the posterior end → increased K^+ conductance of cell membrane → hyperpolarization of membrane → electrotonic spread of hyperpolarization → increased rate of ciliary beat → increased rate of locomotion

In some ciliates, e.g. *Stentor* (Fig. 11), internal conduction of impulses from the bases of cilia is involved (a *neuroid* system). The way in which conduction is achieved in the neuroid system is not well understood. Fibrils beneath the cell membrane connect the bases of cilia, and it was formerly thought that co-ordination of ciliary beat was due to this network. However, there are insufficient cross-connections in this system to explain the way in which the cilia are moved almost like a field of corn being blown in the wind.

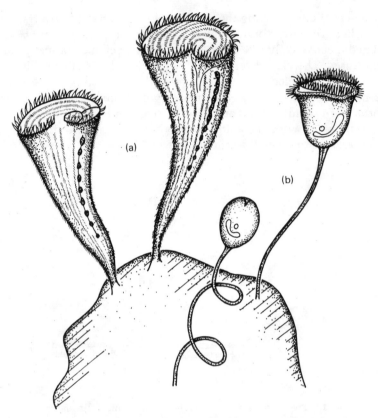

FIG. 11: Ciliated protozoans (AFTER BUCHSBAUM): (a) *Stentor* and
(b) *Vorticella*

In the ciliate *Euplotes*, a system of subsurface fibrils runs from an anterior row of cilia (a membranelle) to a group of fused cilia (cirri). The avoiding reaction of *Euplotes* involves a reversal in beat of the membranelles, and a reversal of beat in the anal cirri. A cut across the fibrils linking the cirri to the membranelles (Fig. 12) results in a loss of co-ordination between these structures, but cuts to other regions of the cell not severing the fibrils have no serious effect on co-ordination. These cutting experiments have recently been repeated, but with conflicting results. Whether the fibrils are concerned with conduction and co-ordination of activities is thus still controversial, and their mode of operation is obscure. Even if the fibrils are important in conduction, this internal system is obviously very different from conduction of impulses in true neurons, where changes occur in the surface membrane.

Some of the behaviour patterns shown by ciliates appear quite variable.

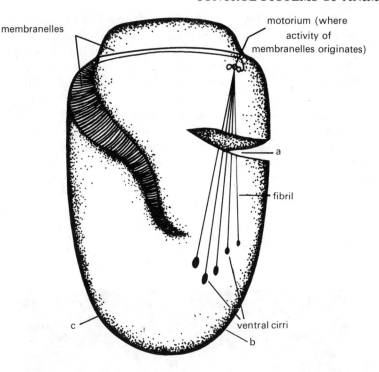

FIG. 12: A cutting experiment testing co-ordination involving internal fibrils of *Euplotes*. A cut (a) through the fibrils running between the membranelles and the ventral cirri results in loss of co-ordination between these structures. Cuts at other points (b and c) do not interfere with co-ordination. (AFTER TAYLOR)

If the fixed form *Stentor* is subjected to continued stimulation by carmine particles which are dropped onto it, it gives a series of responses. The animal will bend from side to side, reverse the beat of its cilia, contract a few times, and if the stimulation still persists, eventually will release its attachment from the substrate and move off.

Most of the work on learning ability of protozoans has been attempted with ciliates, and the results indicate that some learning may occur. However, a number of workers still dispute such claims.

The fixed ciliates *Stentor* and *Vorticella* (Fig. 11) will respond by contraction to a stimulus such as a jet of water or light touch, but after a number of repetitions of the stimulus they fail to respond. This change of response may be considered to occur as a result of 'experience', rather than exhaustion or injurious effects of the repeated stimulation on the animal.

When paramecia were run in a T-maze after a forced turn prior to entering the choice-point, the subjects showed spontaneous alternation, choosing the opposite direction to that of the forced turn. This example of 'reactive inhibition' has been classed as a primitive type of learning, since some form of 'knowledge' of previously executed responses seems to be involved.

Beatrice Gelber trained paramecia to approach a clean platinum wire lowered into a dish containing a culture of the animals. The first presentation of the wire was avoided by the animals; they later 'ignored' it. Training consisted of lightly baiting the wire with bacteria and presenting this for about fifteen trials. The paramecia approached and clung to the wire under these conditions. The culture was tested two to three hours after the training trials by lowering a *clean* wire into the culture. The paramecia approached and clung to the wire. Although these experiments have been criticized on a number of grounds, suitable control experiments appear to have countered these criticisms and perhaps paramecia can be classed as capable of showing complex learning.

SPONGES (PORIFERA)

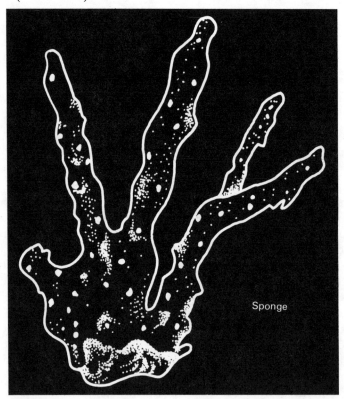

Sponge

These animals are sessile and mainly marine (Fig. 13), but a few freshwater species occur. The sponge body is simple, with numerous small openings (pores) allowing water to enter a central cavity lined with flagellated cells. The water leaves by a larger aperture, the *osculum*. In complex sponges a number of oscula may be present. Although the question is currently a topic of dispute, most workers believe that there are no true nerve cells to be found in sponges.

If the surface of a sponge is stimulated, a localized response is given; oscula or pores may slowly contract. Slow changes in shape are possible in many species. Some of the cells of sponges are spindle-shaped and contractile, and these have been classed as *independent effectors*, since they can operate without nervous control. Contractile cells form a ring around the osculum and are thought to react directly to a stimulus. Closure of the osculum is not correlated with dermal pore closure.

Behavioural responses of sponges are thus very simple; they are limited to such activities as aperture closing, changes of shape and water pumping by the action of flagellated cells.

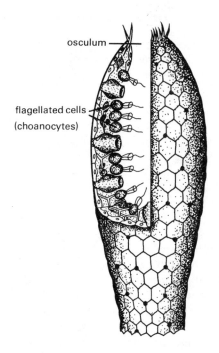

osculum

flagellated cells
(choanocytes)

FIG. 13: A sponge sectioned to show flagellated cells in the central cavity

31

NERVOUS SYSTEMS

The nervous system of multicellular animals (*metazoans*) is usually divisible into two main portions: the *central nervous system* (*CNS*) and the *peripheral nervous system*. Within the CNS a brain is developed, which gives off (in bilateral animals) a longitudinal cord or cords. The major functions of the CNS are to receive and interpret information from outside and inside the body and to initiate nerve impulses which lead to suitable responses by the animal. The peripheral nervous system comprises the nerves which enter or leave the CNS. These nerves are either *sensory* (*afferent*) or *motor* (*efferent*), and their major function is the rapid conduction of impulses. Sensory nerves conduct impulses *from receptors to the CNS*; motor nerves conduct impulses *from the CNS to effectors* (e.g. muscles, glands). Strictly speaking, it is the components of the nerves (the *neurons*) which are termed sensory or motor; thus a nerve may contain only sensory or only motor components, or may contain both (a *mixed* nerve). Peripheral nerves innervate the *viscera* (the gut and associated structures) and circulatory system, etc. (such nerves are termed *visceral*), while *somatic* nerves innervate the muscles involved in locomotion, food collection, etc.

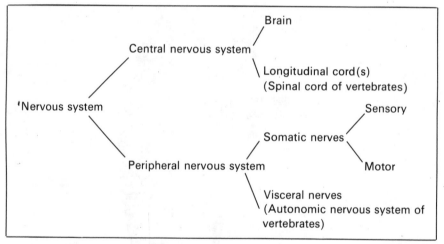

The cells within the nervous system are of two main types, *neurons* (nerve cells proper) and *neuroglia*. Within the CNS neuroglia are often called *glia* cells, and outside the CNS, *Schwann* cells. There are a number of different kinds of neuroglia and their functions are not well understood; they are thought to be involved mainly with structural and nutritive support to the neurons. The neurons can be considered to be the most important cells of the nervous system and are specialized for reception, initiation, conduction and transmission of impulses.

The neuron consists of a cell body (*soma*) with a number of processes—the dendrites and the axon (Fig. 14). Usually the neuron possesses one axon but many short dendrites. The dendrites *receive excitation*, usually from other neurons, while the axon is specialized to *conduct impulses*, sometimes over distances of several metres.

Neurons are commonly classified according to the number of processes they possess, and their function. There may be one, two or many processes arising from the soma; the neurons are then termed *unipolar* (Fig. 15), *bipolar* (Fig. 16) or *multipolar* (see Fig. 14) respectively. Neurons which conduct towards the CNS are sensory, those which conduct away from the CNS are motor, while *interneurons* (connecting neurons) conduct *within* the CNS. A *ganglion* is a region of the nervous system where the cell bodies (somas) of neurons are concentrated.

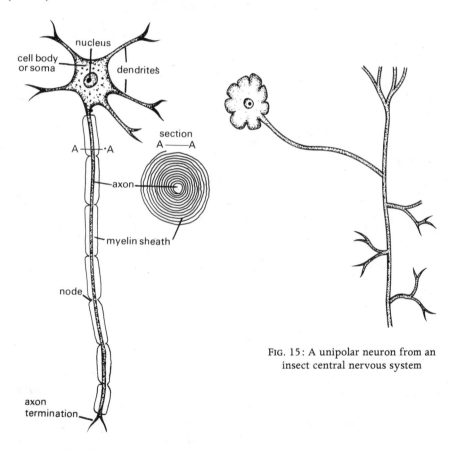

FIG. 15: A unipolar neuron from an insect central nervous system

FIG. 14: A multipolar neuron (AFTER McCASHLAND)

In transverse section the vertebrate spinal cord shows two distinct regions, the *white* and *grey* matter (Fig. 17). The central portion of grey matter is a region where the cell bodies of the interneurons and motor neurons occur. The white matter is a region where axons are cut in section, and the white appearance is due to a sheathing material. The sheath is composed of a fatty material and is termed a *myelin* sheath; the nerve which is so covered is called a *myelinated nerve*. At intervals (*nodes*) the myelin sheath is interrupted (see Fig. 14).

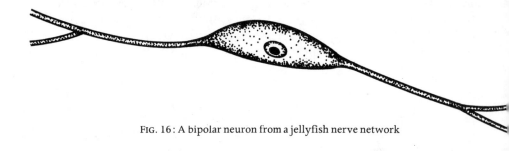

FIG. 16: A bipolar neuron from a jellyfish nerve network

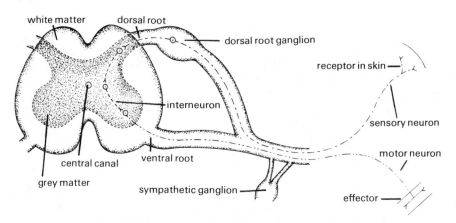

FIG. 17: Transverse section of the vertebrate spinal cord

Presence of a myelin sheath around axons allows impulses to be conducted at a much faster rate than in those neurons where the axons are non-myelinated. In invertebrate animals, speed of conduction has been achieved not by developing a sheath around axons, but by increasing the diameter of the neuron process, producing *giant nerve fibres*. Such fibres are so large that it is relatively easy to insert probes into the axon and record the electrical changes which accompany the passage of impulses. Most of our knowledge of the mode of operation of nerves comes from work carried out on the giant nerve fibres of the squid (Fig. 18).

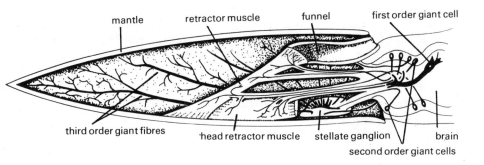

FIG. 18: Giant nerve fibres in a squid. First order giant fibres are restricted to the brain. Second order fibres run from the brain to the stellate ganglion and also to the funnel and neck muscles. Third order fibres run from the stellate ganglion to the muscles in the mantle; those which travel furthest have the largest diameter and thus conduct more rapidly. In this way all the mantle muscles are contracted synchronously and water is directed out of the funnel enabling the animal to move by jet propulsion. (AFTER YOUNG)

In transverse section, the nerve cord of invertebrate animals shows two zones, fibrous and soma-containing as in vertebrates; however, the relationship between them is opposite to that found in vertebrates. Here the nerve cell-bodies occur to the outside of the cord (the *rind*), and the central region (the *core*) consists of fibres (Fig. 19).

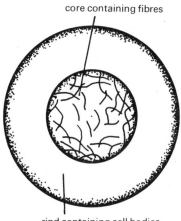

FIG. 19: Diagrammatic transverse section of the cord from the CNS of an invertebrate

35

Neurons are interconnected at special regions called *synapses*. At the synapse, neurons come sufficiently close to one another to allow an impulse to be transmitted from one to the next. The synapse offers a certain resistance to the passage of a nerve impulse; for example, one impulse arriving at the termination of a neuron may not excite the second neuron, while a number of impulses in close succession may be sufficient to excite it. Another important attribute of the synapse is that it allows nerve impulses to pass in one direction only, acting as a one-way valve. The structure of synapses varies greatly, but basically the synapse is a small gap (about 200-300Å)* between the membranes of the impulse-carrying neuron and the neuron to be excited. The relations of the parts and the terminology applied are shown in Fig. 20.

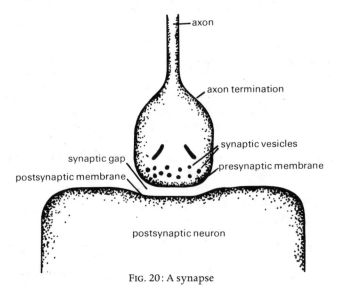

FIG. 20: A synapse

MODE OF ACTION OF NEURONS

All cells are enclosed by a cell membrane, and a potential difference exists between the outside and inside of the membrane, the outside having a positive charge and the inside a negative charge. The cell membranes of metazoans are often about 70 millivolts negative on the inside with respect to the outside.

The reason why such a potential difference exists mainly derives from the permeability of cell membranes and the concentrations of substances on either side of the cell membrane. In metazoans, the tissue fluid which

*$1Å = 1^{-10}m$

36

bathes the cells is usually rich in ions such as sodium (Na^+) and chloride (Cl^-), while the interior of cells is rich in potassium (K^+). The cell membrane is about seventy-five times as permeable to potassium as it is to sodium, and it appears that the basis of the potential difference across membranes is mainly explained by movements of potassium ions.

Consider a living cell (Fig. 21) with no charge on its membrane, but with differing concentrations of ions either side of the membrane. Because the membrane is relatively permeable to potassium, a movement outwards of K^+ occurs by simple diffusion. The loss of positively charged ions in this way and their accumulation on the outside make the inside of the cell membrane negative, and this negative charge immediately tends to attract the positively charged potassium ions back into the cell. Eventually an equilibrium will be reached where loss outwards by diffusion will be balanced by electrochemical attraction inwards. At this equilibrium, the membrane potential is just under $\frac{1}{10}$ volt.

Other ions such as Na^+ also move across the cell membrane. Sodium ions tend to move into cells; they are attracted by the electrochemical gradient, and also diffuse inwards down their concentration gradient. There is no passive gradient tending to move this accumulated Na^+ out of the cell; it is removed by the 'sodium pump', an *active process* which involves the use of energy by the cell.

Excitable cells (nerves and muscles) possess a special facility, that of allowing a very brief alteration in the permeability of the cell membrane, after which the normal permeability relations are restored. The nerve impulse depends on these changes in permeability.

A portion of an axon which is not conducting an impulse has a potential

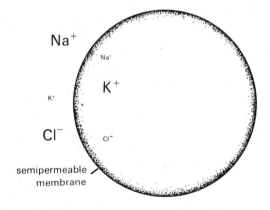

Na⁺

Na·

K⁺

K·

Cl⁻

Cl⁻

semipermeable
membrane

FIG. 21: Relative concentration of various ions on either side of the membrane of a living cell

difference across its membrane, as seen above, called the *resting potential*, and the membrane is said to be *polarized*. When the membrane is stimulated in some way (electrically, mechanically, etc.), the region of membrane receiving stimulation becomes momentarily more permeable to Na^+. Because this ion is highly concentrated outside the cell, the result is a rapid inflowing of Na^+ which results in the inner surface of the axon having, not a negative charge as in the resting condition, but a positive charge; the membrane becomes *depolarized* (whenever the membrane becomes more positive than the resting potential value, the membrane is said to be depolarized). The permeability change is extremely rapid (about 1 milli-second) and immediately afterwards the normal permeability towards Na^+ is restored. Permeability changes also occur with respect to other ions; for example, the potassium permeability is increased, especially immediately after the peak of Na^+ permeability, and this contributes to the rapid return of the membrane to the resting condition. The brief electrical change involved with these ions' movements is called the *action potential*, and the *nerve impulse* refers to the sum of all the changes, i.e. electrical, chemical and membrane permeability changes. When the action potential occurs in a very small region of the axon, it moves rapidly along in both directions from the site of stimulation. The movement is produced by a local flow of current between the inactive and active regions of the axon (Fig. 22), the current being sufficient to alter the permeability of the membrane as it flows through. With the inrushing of Na^+ the membrane becomes de-polarized, and a self-propagating action potential results.

Provided a stimulus is of sufficient intensity to excite the neuron—that is, above the *threshold*—the action potential produced is the maximum that the axon is capable of producing. This is the *all-or-none law*, which states that the action potential is of the same size, irrespective of the magnitude of the stimulus. The action potential is conducted *without decrement* along the length of the axon. Thus the nerve impulse is a brief, discrete event, and because permeability changes (and the flow of ions) occupy a certain (very short) period of time, there is a limit to the rate of conduction of impulses. Because the size of the action potential is indepen-dent of the stimulus intensity, information about the *intensity of stimulation* cannot be sent along neurons in the form of alterations in the magnitude of response. It is accomplished by varying the *frequency* of impulse propaga-tion; low-intensity stimulation results in a low-frequency discharge of impulses along the neuron, the frequency increasing with increasing stimulus intensity.

When the action potential reaches the termination of the axon, the impulse must be transmitted to another neuron (or an effector). Weak electrical signals like the action potential cannot jump the gap of the

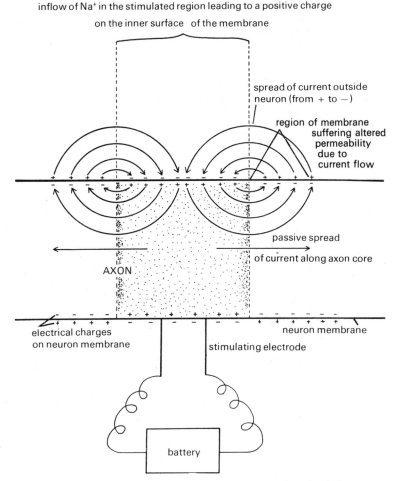

inflow of Na⁺ in the stimulated region leading to a positive charge

on the inner surface of the membrane

spread of current outside
neuron (from + to −)

region of membrane
suffering altered
permeability
due to
current flow

passive spread

of current along axon core

AXON

electrical charges
on neuron membrane

neuron membrane

stimulating electrode

battery

FIG. 22: Electric charges and currents in a nerve under stimulation

synapse, so that chemical means are employed to excite the postsynaptic neuron. (Electrically-transmitting synapses are known, but are rare. Their synaptic gap is very narrow.) The synaptic vesicles at the axon termination (see Fig. 20) contain chemicals known as transmitter substances. Acetylcholine and adrenalin are common transmitter substances in vertebrate nervous systems. On receipt of a nerve impulse, a number of synaptic vesicles release their transmitter substance and this chemical diffuses across the synaptic gap and influences the postsynaptic membrane. If sufficient transmitter substance is released, the postsynaptic membrane is excited and produces an action potential.

The action of the transmitter substance on the postsynaptic membrane causes either *depolarization* (a brief decrease of membrane potential) or *hyperpolarization* (a brief increase of membrane potential). Depolarization is caused by an *excitatory* stimulus, and hyperpolarization by an *inhibitory* stimulus.

With an excitatory stimulus, the postsynaptic response (depolarization) is termed an *excitatory postsynaptic potential (EPSP)*. Release of a small amount of transmitter substance results in only a small EPSP, and the postsynaptic neuron does not propagate an action potential. With the release of a larger amount of transmitter substance, the larger EPSP produced may be sufficient to initiate an action potential which is propagated along the axon. Thus the initial response of the postsynaptic membrane is a *graded* one and the small depolarizations are *not propagated*, although they do spread passively with rapidly diminishing magnitude to adjacent regions of the same neuron. This is in contrast to the *all-or-none*, *propagated* action potential that is characteristic of conduction in the axon.

The release of a certain amount of transmitter substance following each impulse arriving at the axon terminal can explain *summation* at the synapse. A single impulse at the presynaptic membrane may not lead to the production of an action potential in the postsynaptic membrane, but two or more impulses in close succession may be sufficient. This is termed *temporal* summation, since the impulses are separate in time. *Spatial* summation may occur when a neuron, which has many axon terminations on its dendrites or soma, receives excitation via two or more separate axon terminations.

Not all synapses are *excitatory* in nature; some are *inhibitory*. When an inhibitory synapse is operating, the effect tends to suppress the production of an action potential by the postsynaptic neuron, even though it may be receiving impulses from excitatory synapses at the same time. With an inhibitory stimulus, the postsynaptic response (hyperpolarization) is termed an *inhibitory postsynaptic potential (IPSP)*, and it tends to cancel out depolarizations caused by excitatory stimuli.

The dendritic and soma membranes of the postsynaptic neuron can deal with any number of excitatory or inhibitory influences, provided their sum does not exceed the threshold level for the neuron. When the threshold level is reached, a propagated action potential results. The dendritic and soma membranes, therefore, are regions of the nervous system where *integration* occurs. A large proportion of the surface area of these membranes receives synaptic contacts from other neurons. The membranes may simultaneously receive information via many synapses, and their special properties of *passive* spread of potential and *graded* responses allow integration to occur at sub-threshold levels of membrane potential. Excitatory and inhibitory potential changes are summed, and if the result is a sufficiently large depolarization, impulses of a certain frequency leave the neuron via the axon.

The properties of the nervous system govern the expression of an animal's behaviour, and many examples can be given to demonstrate the concepts of summation, inhibition and other related physiological changes. Blowflies 'taste' food using receptors in their fore-legs, and when suitably stimulated the flies push out their proboscis to feed. Under experimental conditions it was found that dipping one leg into a very weakly concentrated sugar solution did not lead to proboscis extension, but dipping *both* legs into the same solution did. The stimuli had been summed spatially.

Inhibitory action is commonly involved in the activities of the nervous system. In most activities, such as walking, the action of some muscles is inhibited while their antagonistic muscles are operating. In general, the brain of insects can be said to exert an inhibitory effect. Lower motor centres in the CNS, such as the ganglia controlling walking, are in turn controlled by the brain, which can remove the inhibition from these parts to elicit the required type of behaviour.

One of the best-known cases of inhibition of a motor centre in insects involves the reproductive activities of the preying mantis. Impulses producing copulatory movements by males would be initiated continuously by the last abdominal ganglion, if it were not for the inhibitory control exerted by the suboesophageal ganglion. On contact with the female, the inhibitory effect is removed, leading to copulation.

Mantids are cannibalistic and the males may be attacked by the females. Since the anterior regions are the most vulnerable to attack, males often literally lose their heads to the females. The result, as far as successful mating is concerned, is beneficial, since the suboesophageal ganglion is frequently removed and with it any inhibition on the copulatory centre; the activities of the male abdomen are carried out with more vigor than when the body was intact.

RECEPTORS

Survival demands that most of the changes in an animal's external and internal environment are detected and an appropriate response made. Fish living in many of our rivers need to move away from areas of high temperatures or concentrations of industrial wastes. The carbon dioxide concentration of our blood is continually monitored; any increase leads to an increase in our breathing rate or a decrease in the rate of oxygen consumption. Receptor cells detect such environmental changes.

All cells in the body are sensitive, or show *irritability* and can be stimulated to change their activities. In *Protozoa* the whole body acts as a receptor cell and can detect a variety of stimuli. In *Metazoa*, since many cells are present, some can be specialized to respond predominantly to one type of stimulus (the *adequate* stimulus), but they still retain their general excitability and can respond to other types of stimuli (*inadequate* stimuli). Receptor cells in the eye are particularly sensitive to light, but they also respond to mechanical pressure; for example, we 'see stars' after receiving a blow to the eye.

Animals detect three parameters of the stimulus they receive: (a) its nature, (b) its intensity, and (c) its position.

DETECTION OF THE NATURE OF STIMULI

The *nature* of different stimuli can be detected by the use of different types of receptors. Thus the following receptors can collect corresponding information about the physical and chemical properties of the environment.

RECEPTORS	STIMULUS TYPE
Chemoreceptors	chemicals
Photoreceptors	light
Thermoreceptors	temperature
Electroreceptors	electric current
Mechanoreceptors	pressure,
	touch,
	air vibrations,
	stretch,
	sounds,
	gravity,
	acceleration

These receptors not only differ in the type of stimulation to which they respond, but are different morphologically (Figs. 23 and 24, Plate II). However, there are some tissues, like the cornea of the eye, which contain only free nerve endings as receptors, yet are sensitive to touch, temperature and noxious stimuli.

DETECTION OF STIMULUS INTENSITY

The *intensity* of a stimulus received by a receptor cell is coded in nerve impulses. All receptors function in the same general way: they act as transducers, converting one form of energy into another. The various stimuli impinging on the animal are transduced into the 'language' of the nervous system, a train of nerve impulses of a definite frequency.

The neurophysiological events, from the receipt of stimulus energy to

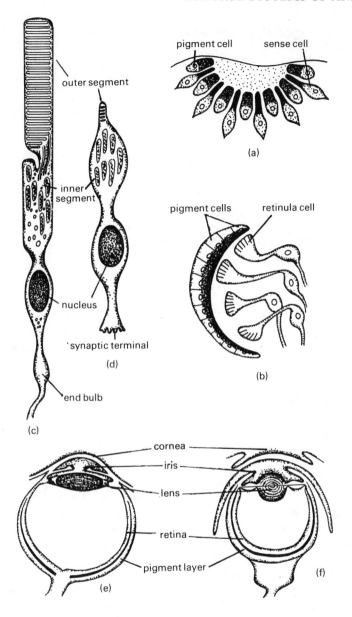

FIG. 23: Some photoreceptors of animals: light-sensitive cells of
(a) a coelenterate (AFTER RAMSAY) and (b) *Planaria* (FROM HYMAN);
(c) rod and (d) cone in a vertebrate retina (MODIFIED FROM WOOD);
and (e) vertebrate, and (f) cephalopod, eyes (AFTER RAMSAY)

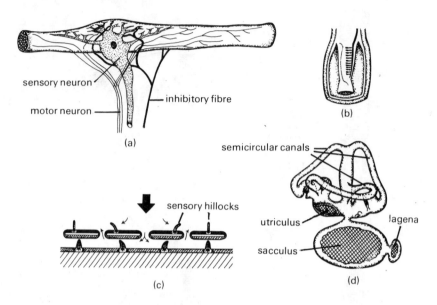

FIG. 24: Some mechanoreceptors of animals (FROM FLOREY): (a) a stretch receptor of crayfish, (b) tympanic organ of a grasshopper, (c) lateral line organs of fish under stimulation by the pressure and current (the big arrow) from an approaching object causing fluid movement (directions indicated by small arrows) through canal pores, and (d) a labyrinth of fish

the passage of impulses towards the CNS, are outlined diagrammatically in Fig. 25. Absorption of stimulus energy leads to depolarization of the receptor cell membrane, termed the *receptor potential*. There are two important differences between receptor potentials and the action potentials (nerve impulses) which occur in nerves. Action potentials are all-or-none responses and are self-propagating (see previous section). Receptor potentials are graded responses, and the amount of depolarization is related *logarithmically* to the stimulus intensity. The depolarization spreads passively over the cell membrane and rapidly diminishes with distance (i.e. is non-propagated). Once the receptor potential reaches or exceeds a particular threshold value, a train of nerve impulses is generated (all-or-none, propagated responses), and the frequency with which they leave the receptor site is related *linearly* to the amount of depolarization. Transduction by the receptor is thus a two-stage process: in the first stage, stimulus energy is coded into an electrical potential change whose amplitude is related to stimulus intensity; in the second stage, this receptor potential is coded into a sequence of nerve impulses whose frequency is related to the amount of depolarization.

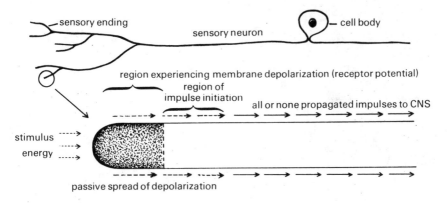

FIG. 25: Diagrammatic receptor neuron with a sensory ending enlarged to indicate some events following stimulation

Although most receptors function in the way outlined above, intermediate processes sometimes occur, as in photoreceptors and some non-nervous sense cells. Light-sensitive cells in both vertebrates and invertebrates contain a photosensitive pigment, and the first stage in the process which leads to the production of a receptor potential involves the absorption of light by the pigment, and subsequent photochemical changes. In vertebrates, the taste buds (Fig. 26) and also the hair cells of the ear are specialized non-nervous cells and the initial transduction occurs here, giving rise to a generator potential. The receptor potential arises in the membrane of the sensory neuron closely applied to the non-nervous cell.

Impulses initiated by the receptor cell membrane leave at frequencies varying from about 1 to 1000 per second. One thousand impulses per second is near the upper limit at which impulses can be carried along neurons. Frequencies lower than 1 per second are of course possible, but when the frequency of impulse production falls much lower than this, the information coded by the receptor is liable to misinterpretation. Since there may be background 'noise' to the cells of the nervous system, 'stray' or spontaneous nerve impulses would be interpreted as meaningful stimuli.

The range of impulse frequencies available for use by a receptor to code intensity of stimulation is thus about 10^4, but the range of intensity presented to a sense cell may be extremely large, e.g. more than 10^8 in photoreceptors. Some values of natural light intensities which we, using our photoreceptors, receive and can code into appropriate impulse frequencies are indicated in Fig. 27.

The problem of coding the output of a receptor to cover a large stimulus intensity range is solved in two main ways. Firstly, since the receptor potential varies as the logarithm of the stimulus intensity, a large change in

45

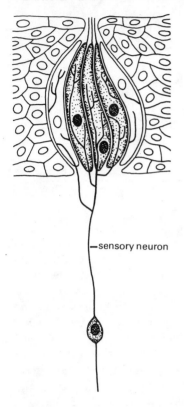

—sensory neuron

FIG. 26: Taste receptors of the tongue as non-nervous receptor cells. Cell bodies are buried under the surface. (AFTER LOEWENSTEIN)

intensity at high stimulus intensity levels brings about the same amount of depolarization of the receptor cell membrane that only a slight change in levels would produce at low light intensities. Secondly, receptor organs in animals rarely consist of only one receptor cell; usually many cells are present. Often the sensitivities of the receptor cells vary, allowing the range of stimulus intensities capable of being signalled to be greatly expanded compared with a single receptor. The rod cells in the retina of the eye are specialized to operate at low light intensities; in bright light they are relatively ineffective, but after about half an hour in the dark they attain their maximum sensitivity. The cone cells of the retina function efficiently in bright light, and the combination of rods and cones thus allows the eye to operate over a very wide intensity range (at least eight logarithmic units). The cones also differ amongst themselves, being sensitive to light of different wavelengths, and this gives us the basis for our colour vision.

The output of most receptor cells depends not only on the stimulus intensity but also on the time since the stimulus was applied. Fast-adapting or *phasic* receptors have a high output frequency when the stimulus is first

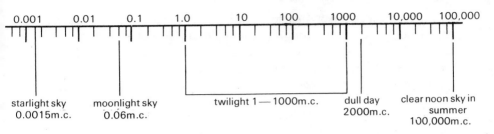

FIG. 27: Some values of natural light intensities (metre-candles) (AFTER HARDEN JONES)

applied, but the frequency rapidly diminishes and finally ceases altogether (Fig. 28). Touch and temperature receptors are phasic, and this means that the CNS is not being continually supplied with information about events occurring on the surface of the animal unless the stimuli arriving there change. Slow-adapting or *tonic* receptors maintain their output at a steady frequency for long periods. Mammalian receptors detecting blood pressure are tonic; they continue to initiate impulses which allow the animal to maintain a stable blood pressure. Receptors concerned with posture are largely tonic, otherwise we would not be able to stand up or walk without conscious effort.

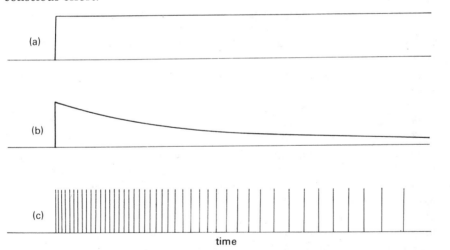

FIG. 28: (a) The stimulus intensity, (b) sensation experienced, and (c) nerve impulses leaving the receptor, when a phasic receptor is continuously stimulated (AFTER ADRIAN)

DETECTION OF THE POSITION OF STIMULI

The *position* of a stimulus relative to the body is indicated by the position of the particular receptor involved. One method of receptor classification

47

is based on the location of their stimulation, allowing us to recognize distance receptors, exteroceptors, proprioceptors and interoceptors.

RECEPTORS	LOCATION AND TYPE OF STIMULUS
Distance receptors	events which actually occur at a distance from the body, but the stimuli are received at the body surface (light, odour, noise)
Exteroceptors	events occurring on the body surface (touch, pressure, taste)
Proprioceptors	situated to detect the positions and attitudes of the body and its parts, e.g. stretch receptors in muscles and tendons (stress, strain, pressure)
Interoceptors	visceral receptors concerned with internal regulation (pressure and temperature of blood, visceral pain)

EFFECTORS

Effectors allow animals to react to their environment, and the commonest way in which they do this is to use muscles. Locomotion or movement of the body's parts is brought about by the contraction of thousands of fine protein filaments within each muscle cell.

Muscles responsible for locomotion are often inserted onto parts of the skeleton which function as a system of levers. In hollow structures like the gut, the muscles are inserted onto other muscles, not to any skeletal structure. Movement of materials within the gut is produced by the waves of contraction which pass along its walls.

Protozoans, which lack muscles, use a variety of methods to bring about movements. Protoplasmic projections may be put out, the rest of the body flowing after them, as in Amoeba; contractile threads in the stalks of forms like Vorticella allow retraction; ciliates and flagellates possess short, hair-like cilia or flagella whose beating drives the animal through the water. In the cycle of ciliary beat (Fig. 29), the cilium is held relatively rigid during the propulsive down-stroke, but is more limp during the recovery stroke. Protozoans are able to reverse the direction of the active stroke, while this is not possible in the cilia on mammalian tissue. When mammalian sperms are moving, the movement of the flagellum is usually undulatory and waves of contraction pass from the base toward the flagellum tip.

Some effectors such as glands, chromatophores and electric organs are not concerned with movement.

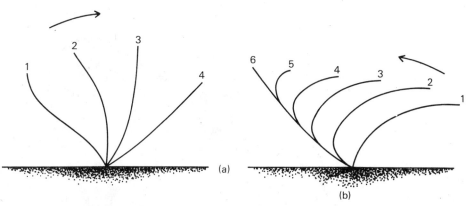

FIG. 29: Ciliary beat of *Paramecium* (AFTER SLEIGH): (a) propulsive stroke and (b) recovery stroke

When effectors first appeared in the animal kingdom they probably operated directly in response to their environment; nerve cells were lacking at this stage. Effectors which operate in this way and whose activities are not initiated by neurons or receptors are retained by some animal groups, and are called *independent effectors*.

The sponges, considered by most zoologists to lack nerve cells, can close their exhalent aperture, the osculum, in response to touch, injury, carbon dioxide, lack of oxygen, exposure to air, harmful chemicals and sometimes bright light. The effector involved is a loose ring of elongated contractile cells around the osculum.

Stings from the sea wasp or box jellyfish *Chironex fleckeri* (Fig. 30) can be fatal to a human within minutes. However, the toxin from coelenterates is usually not harmful to man. Coelenterate stinging cells are classic examples of independent effectors. Mechanical and chemical stimuli (e.g. contact with prey) are detected by a trigger-like structure. This leads to the explosive eversion of a coiled-up thread within the *nematocyst* capsule (Fig. 31). The threads are used for prey capture, defence or attachment. They may be adhesive, coiled for wrapping around objects, or penetrating. The latter type introduces the toxin into the prey—or any other object that has caused nematocyst discharge.

The cilia and flagella of protozoans are obviously operating independently of nervous influence. The cilia of metazoans are most commonly found lining cavities such as the fallopian tubes down which eggs travel after release from the ovary. In such a situation, cilia may operate spontaneously and with independent action. However, in some sites, nervous influence may be superimposed on spontaneous activities to regulate them.

FIG. 30: The sea wasp *Chironex fleckeri*, a box jellyfish. Stings from this animal in Australian waters have caused several fatalities.

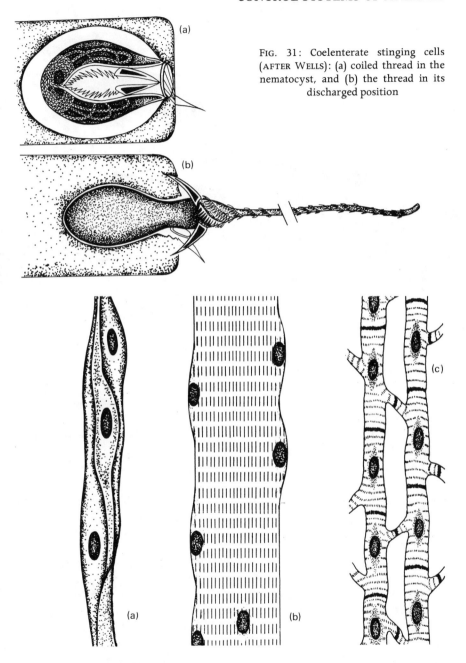

FIG. 31: Coelenterate stinging cells (AFTER WELLS): (a) coiled thread in the nematocyst, and (b) the thread in its discharged position

FIG. 32: Different types of muscle: (a) visceral (smooth), (b) skeletal (striated), and (c) cardiac

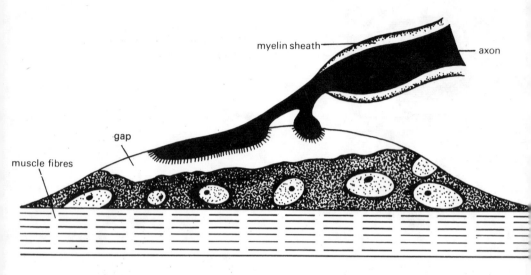

myelin sheath

axon

gap

muscle fibres

FIG. 33: A motor end-plate (MODIFIED FROM COUTEAUX)

We have seen that locomotion in protozoans may involve a number of different structures, but in metazoans only one type of tissue is used—muscular tissue. All muscle cells depend upon fibrillar proteins to bring about their contraction, but the internal organization of the muscle cell differs in different muscle types (Fig. 32). Muscles are termed striated or smooth, depending on the presence or absence of cross-striations. Heart (cardiac) muscle and those attached to our skeleton (skeletal muscles) are striated, while the muscles of the gut (visceral muscles) are smooth. Although their activities are regulated by the nervous system, smooth and cardiac muscles contract and relax rhythmically, even when their nerve supply has been cut. Striated muscles, however, are under direct nervous control; if their nerve supply is cut they no longer function.

Muscles are innervated by motor neurons which terminate at the *motor end-plate* or *neuromuscular junction* (Fig. 33). There is a small gap between the membranes of the neuron and the muscle cell. Chemicals are used to transmit excitation from the neuron to the muscle, in much the same way as the neuronal synapse is bridged. The transmitter chemicals cause depolarization of the muscle membrane, and the effect spreads inwards causing the protein filaments to contract. The whole muscle then shortens and develops tension.

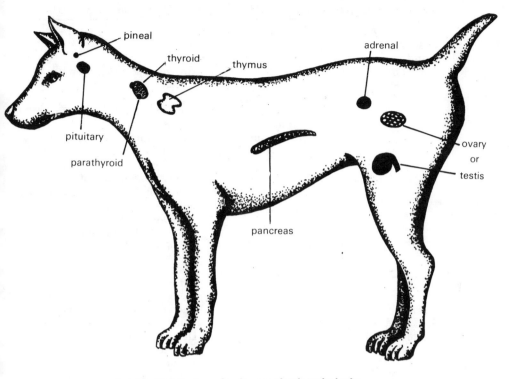

FIG. 34: Distribution of endocrine glands in the body

ENDOCRINE SYSTEMS

The fact that behaviour is controlled by the nervous system has been known from ancient times, but only during the late nineteenth and early twentieth centuries has the role of the *endocrine system* been appreciated.

Within the cells of the body, numerous chemical reactions are occurring. The function of the endocrine system is to co-ordinate and control these metabolic activities so that the body operates as an integrated unit and not as a system of independent and unrelated reactions.

Endocrine glands are found in various parts of the body (Fig. 34). Their secretions, called *hormones*, are passed directly into the bloodstream because the glands have no ducts. Endocrine glands are often called ductless glands. The majority of endocrine glands are found in vertebrates.

53

ENDOCRINE GLANDS OF VERTEBRATES *(particularly mammals)*

GLANDS	HORMONES	FUNCTIONS
Pituitary gland		
Anterior lobe	Growth hormone	Growth and metabolism in general
	Gonadotropic hormones:	
	follicle-stimulating hormone (FSH)	Growth and development of gametes
	luteinizing hormone (LH)	Final maturation of gametes; control of secretion of the male hormone *testosterone* from interstitial cells of the testes in the male; ovulation of mature follicles and formation of *corpus luteum* in the female
	Thyrotrophic (thyroid-stimulating) hormone (TSH)	Growth of the thyroid and control of its secretion (*thyroxin*)
	Adrenocorticotrophic hormone (ACTH)	Stimulation of adrenal cortex for production of other hormones
	Lactogenic hormone	Secretion of milk from the mammary glands
Intermediate lobe	Melanocyte-stimulating hormone	Darkening of the skin in lower vertebrates by influencing pigment cells (e.g. melanophores of the frog skin)
Posterior lobe	Vasopressin	Increase in the absorption of water in the kidney tubules and contraction of peripheral blood vessels
	Oxytocin	Contraction of the uterus; the 'let down' of milk by mammary glands
Adrenal glands		
Medulla	Catecholamines:	[Inhibition and excitation of various muscles for fighting or fleeing]
	adrenalin	Blood flow to the digestive tract and other organs not required in the emergency state
	noradrenalin	Decrease in blood flow through the brain, and increase in the total resistance of peripheral organs with increase in blood pressure in the body

Cortex	Corticosteroids (many hormones)	Electrolytic balance; water metabolism; carbohydrate, fat and protein metabolism; control of blood volume and pressure; interaction with gonads; anti-inflammatory effects
Gonads		[Development of secondary sexual characters in lower vertebrates]
Testis	Testosterone	Development of accessory genitalia and secondary sexual characters, and interaction with the anterior pituitary for reduction in LH
Ovary	Estrogen	Development of the uterus, vagina, mammary glands and other secondary sexual characters, and interaction with the anterior pituitary for reduction in LH
	Progesterone (from corpus luteum)	Enlargement of the uterus, development of the placenta and enlargement of the mammary glands
Thyroid gland	Thyroxin and triiodothyroxin	General cellular respiration; control of metamorphosis in Amphibia
Parathyroid glands	Parathormone	Calcium metabolism
Pancreas	Insulin	Transportation and oxidation of sugar in cells; storage of glucose as glycogen; conversion of excess sugar to fat; production of glucose from other substances
	Glucagon	Interaction with the adrenal cortex and pituitary hormones and increase of blood sugar (opposite effect to insulin)
Thymus	Thymic hormone	Increase in resistance to infections and immunity reactions
Pineal gland	?	Inhibitory effect on gonadotropic secretion?

Hormones usually exert their effect on some part of the body far from the site of the gland which produced them. Every cell of the body, nourished by the bloodstream, is exposed to the influence of hormones. However, only a few types of cell 'recognize' a particular hormone. Within these target cells the *rate* of certain chemical reactions is altered; hormones do not *initiate* reactions. Specificity of action thus depends on the type of cell which responds to the distributed 'chemical message'.

A hormone from one species can often produce its specific effect when introduced into the bloodstream of a different species; diabetics in need of the hormone *insulin* may use that extracted from cattle.

Control of chemical activities is often demanded for long periods, and the endocrine system can meet such demands by maintaining a steady secretion for as long as is necessary. A woman, when pregnant, must not lose the lining of the uterus in menstruation, for if this should happen the developing embryo would be lost. The wall of the uterus is maintained for the vital nine months by the continued secretion of the hormone *progesterone* from the ovary and the developing placenta.

Normally, when females are not carrying embryos, progesterone is secreted cyclically. For fourteen days after menstruation, progesterone levels are very low. Following the release of an egg or eggs at ovulation, progesterone production begins and the level increases for the next fourteen days, only to fall again after menstruation. At least three other cyclically produced hormones are involved in the control of the menstrual cycle.

The nervous system, which relies in its functioning on the maintenance of a train of impulses to an effector, is much less suited to this type of prolonged control. A further difference between nervous and endocrine systems is seen in the duration of the *effects* produced by both. The effect of nervous stimulation ceases immediately after the passage of impulses, whereas a hormone will linger and continue to influence its target cells for some time after the secretion has stopped.

Because hormones are transported by the bloodstream (or by the tissue fluids if the animal lacks a circulatory system), responses initiated by the endocrine system are only executed slowly. One of the most rapid hormone-controlled responses is the colour change of chameleons. When placed in a new situation they alter their colouration, often matching the background. The change occurs within a few minutes, but may take over half an hour to complete. In the octopus and cuttlefish, where colour change is under direct nervous control, complete changes take less than a second.

Endocrinologists have concentrated most of their attention on vertebrates; invertebrates are usually small, so that their endocrine glands or tissues are small and often difficult to locate. Consequently much less is known about the role of hormones in invertebrates. However, very simple experimental techniques can sometimes be used to demonstrate hormonally-controlled reactions, such as colour change in stick insects.

The stick insect *Carausius morosus* is usually light-coloured during the day and dark at night. Factors such as light, temperature and especially humidity affect the distribution of pigment in the skin cells of this insect.

H. Giersberg tied a ligature around the thorax of a light-coloured stick insect and placed the hind end of its body in a moist box (Fig. 35). Only that part of the body anterior to the ligature became dark, even though this part was not experiencing high humidity. The ligature prevented body fluids circulating past that point, but did not prevent the passage of nerve impulses along the cords. It was concluded that the posterior, stimulated part of the body sent information to the brain via the nerve cords. The brain stimulated release of its pigment-dispersing hormone, but since this chemical could not pass the ligatured region, only the anterior parts darkened. When light-coloured insects without ligatures were placed in the same experimental situation, the whole body became dark in response to the high humidity.

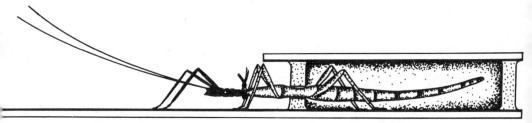

Fig. 35: Giersberg demonstrated hormonal control of colour change in the stick insect *Carausius morosus* by enclosing the rear end of the animal in a moist chamber and placing a ligature between the first and second thoracic segments. The ligature did not interrupt the nervous pathways that carried impulses from abdominal sense organs to the brain but prevented blood flow, so that the colour change hormone secreted under the control of the brain did not reach the regions behind the ligature. When the ligature was removed the whole animal darkened.

The outward expression of behaviour can be very dependent on the particular hormone or hormones being secreted at that time. The larvae of the hawk moth *Mimas tiliae* may move either up or down in relation to gravity, and this geotactic behaviour is influenced by the concentration of juvenile hormone. The larvae feed and moult on twigs and branches of the linden tree. But the larval stage just before pupation (the last larval instar) becomes positively geotactic; the larvae thus move down the tree and dig into the ground where pupation occurs. Experimental implantation of active endocrine glands producing juvenile hormone into last instar larvae produced larval-pupal intermediates. Those which were more pupal than larval were positively geotactic and moved downwards; those which were more larval than pupal were negatively geotactic and stayed at the top of the tree. Positive or negative responses to gravity were thus dependent on the amount of juvenile hormone present.

At least three hormones are involved in controlling the developmental stages (egg → larva → pupa → adult) of the hawk moth (this system is basically the same for all insects). Brain hormone is passed to a pair of storage organs, the *corpora cardiaca*, situated close to the brain (Fig. 36). The brain hormone is transported by the blood and triggers activity in the prothoracic gland, which in turn produces its moulting hormone, *ecdysone*. This hormone activates the cells of the skin, and preparation for moulting begins. Juvenile hormone is produced by the *corpora allata*, a pair of glands close to the corpora cardiaca.

When both ecdysone and juvenile hormone are present, moulting is followed by another larval stage. When reduced amounts of juvenile hormone are available, a pupa is formed. In the absence of this hormone, i.e. when ecdysone is acting alone, adults are produced. Removal of corpora allata, and thus the juvenile hormone, from early instar larvae results in

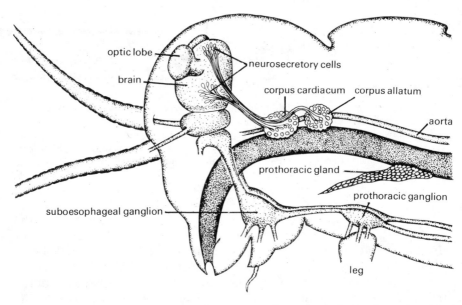

FIG. 36: Endocrine organs in an insect. The corpora cardiaca near the brain release a hormone which stimulates the prothoracic gland to produce its hormone, ecdysone. This latter hormone initiates moulting. Juvenile hormone is secreted by the corpora allata in diminishing amounts throughout larval life. When it is finally absent, the next moult is into the adult phase. (AFTER WELLS)

premature pupation and miniature adults. Giant larvae can be produced by implanting active corpora allata into last instar larvae.

An hormonal change in springtime, influenced by increasing day length, leads to migratory behaviour by males of the three-spined stickleback, *Gasterosteus aculeatus*. They swim upstream to suitable breeding sites and establish territories prior to nest-building and mating. Similarly, behavioural responses concerned with the reproductive activities of many other animals, especially vertebrates, can only be elicited at certain times, i.e. when the appropriate hormone influencing reproductive acts is present in sufficient quantity.

Injection of hormones into different regions of the brain can tell us if a particular hormone influences specific behaviour patterns. When *testosterone*, the hormone from the testes, was injected into a region near the hypothalamus of a male rat's brain, male sexual activities were shown. But when the hormone was delivered to a slightly different brain region, *maternal* behaviour was triggered; the male built a nest from available materials, and baby rats were carried to the safety of the nest. A 'normal' male would eat the young in similar circumstances. To explain the female-type behaviour it was suggested that certain cells in the brain are stimulated by male or female gonadial hormones (from the testis or ovary), and in these experiments sufficient quantities of male hormone to mimic the effect of female hormone were injected, thus affecting the site controlling female-type behaviour. Experiments such as these are beginning to shed light on a topic about which we currently know very little—how hormones may influence and integrate the activities of nerve cells.

By radically altering the rates of certain reactions, hormones can lead to the accumulation or depletion of important body requirements, which in turn can influence behaviour. The following is an example of this.

There was a lecturer who was well liked and respected by his students. It was all the more surprising therefore to see such a 'steady' individual apparently depressed and drunk one day in the presence of his class, and using language hardly fit for his assembled audience. In fact, excessive intake of alcohol was not the reason for the unusual display. His blood glucose level was well below normal, and were it not for the sugar administered by a student who realized his condition a coma would have resulted. The patient was suffering from sugar diabetes (diabetes mellitus). His pancreas was producing insufficient insulin, a hormone which causes lowering of the blood glucose level. Treatment for this condition, an unduly high blood glucose level, involves regular insulin injections. But occasionally, if the patient under such treatment has skipped a meal or been subjected to physical exertion, the blood glucose level may drop below normal and unusual behaviour results.

ust seen that hormones can have important effects on behaviour. hand, behaviour can sometimes influence hormone production een in the ring dove (p. 12). In some mammals, such as the cat ovulation needs to be triggered; it is not a naturally occurring part of the oestrous cycle. The act of copulation triggers a nervous reflex linking the hypothalamus in the brain. As a result, a hormone is secreted in larger amounts and initiates ovulation.

PHEROMONES

Hormones are usually transported to their target organs by the bloodstream. A class of chemical substances which function in a somewhat similar way to hormones but which are secreted to the *outside* of the body are called *pheromones*. A pheromone from an animal affects other members of the same species which receive it.

Wherever animals communicate with each other by chemical means, pheromones are involved. The use of chemical signals to achieve sexual attraction and sex recognition is widespread in the animal kingdom; examples can be quoted from all animal groups. A bitch is periodically seen to attract many of the neighbourhood's dogs. She is on heat and about to ovulate at this time, and a pheromone is released into the urine.

A pheromone may only be effective over short distances—about 1 mm in stalked ciliates like *Vorticella* (see Fig. 11, p. 28), where in sexual reproduction one member becomes motile and is chemically attracted to the stalked individual releasing the pheromone. At the other end of the scale, the large feather-like antennae of the male gypsy moth can detect the odour from the female scent glands over distances of 4 to 5 miles.

Chemical alarm signals are perhaps best developed in the social insects such as ants and bees. Alarmed ants push one another and also release pheromones from *anal glands* associated with the stinging apparatus (Fig. 37) or from *mandibular glands* located near the mouthparts. This signal often releases attack in large, powerful ants, but smaller species such as *Lasius niger* (Fig. 38) are stimulated to run away.

Honey bees release an alarm substance in the act of stinging. If an intruder comes close to the hive, the disturbed bee circulates at the hive entrance with abdomen raised and the groove of its sting exposed (Fig. 39). The pheromone alerts many other bees to come out of the hive and attack. Stinging is not the only method of attacking intruders; some bees bite, while others keep a stock of sticky resin within their hive specifically for smearing over their victims.

Pheromones are used quite extensively by many species of ants and bees for making trails. These are used by other workers to locate rich sources of

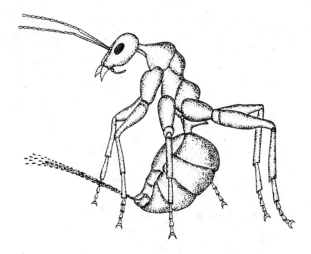

FIG. 37: An alarmed worker of the ant *Formica polyctena* ejecting a fluid containing poison and an ant-alerting substance (AFTER MASCHWITZ)

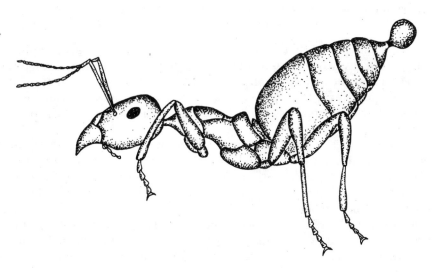

FIG. 38: The ant *Lasius niger* exposing an odorous droplet from the anal region. Fellow ants flee from such a chemical signal. (AFTER MASCHWITZ)

61

FIG. 39: A disturbed honey bee releases its alerting chemical while exposing the sting and stimulates other bees to attack (AFTER MASCHWITZ)

food or sometimes to find their way back to the nest. The effect of a phero-mone is not always a 'releaser', a direct triggering of a behavioural response. Some pheromones alter the physiology of the recipient, and in turn, the physiological changes may lead to an alteration in the behaviour of the animal. Such effects class the pheromone as a 'primer' rather than a 'releaser'.

Sex inversion occurs in the slipper limpet *Crepidula fornicata*. Normally, young individuals are males and later they transform into females. Good places to collect these limpets are oyster beds, where they can be serious pests. Groups of up to twelve individuals form a chain and each clings to the shell of the one beneath (Fig. 40). The chain bends to the right: all right rims of the shells are close together, and because the reproductive openings occur on the right side of the body these too are approximated. This is important, because only the topmost individuals of the chain are able to move.

Grouping of individuals has certain advantages, one of which is the more powerful feeding currents that a number of individuals can produce.

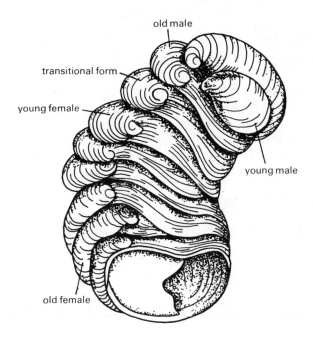

old male

transitional form

young female

young male

old female

FIG. 40: A chain of slipper limpets *Crepidula fornicata*. Sex change occurs in these molluscs. Individuals at the top of the chain are males; those at the base are females, while those in between are in transition from male to female. Immature individuals become males which transform rapidly into females unless they are exposed to a pheromone produced by the female. A young limpet settling at the top of such a chain thus has a prolonged male existence.

(AFTER FRETTER AND GRAHAM)

Slipper limpets are filter feeders and strain fine food particles out of the water current which passes over their gills.

The youngest and smallest individuals in the chain are males; the oldest and largest individuals at the base of the chain are females, while those in between are in transition from the male to the female condition (Fig. 40). Females produce and release into the water a pheromone controlling development of the male phase. An immature limpet settling on the top of the chain develops into a male and has a prolonged male existence. A youngster kept in an aquarium with a female, but out of physical contact with her, develops similarly. However, if the young limpet settles in the absence of a female, the male phase is usually very brief and is followed by a rapid transition into the female condition. Sometimes such an immature limpet can pass directly from the non-sexual to the female condition.

63

Pheromones from *Crepidula* females control the development of a sexual phase, whereas those produced by the queen of a honey bee colony actually inhibit the development of ovaries in the female workers. Workers attending the queen lick her, and in this way pick up 'queen substance' (9-oxo-2-decenoic acid) produced in her mandibular glands. Every worker obtains some of this substance through the mutual food exchanges occurring in the hive. If the queen dies, if workers are separated from their hive-mates by a double wire screen preventing food exchanges, or if numbers in the hive become too large, all or some of the workers are deprived of the substance. Ovaries may be developed by some workers, but more importantly, changes take place in the type of cells being built. Queen cells, larger than cells in which worker larvae develop, are now produced and new queens are formed.

Thus the effect of a pheromone may be immediate (causing a behavioural change) or delayed (the pheromone affecting a developmental process such as maturation or sex determination). Pheromones are used in communication and are better developed in social animals.

4 Evolution of Control Systems (I)—Invertebrates

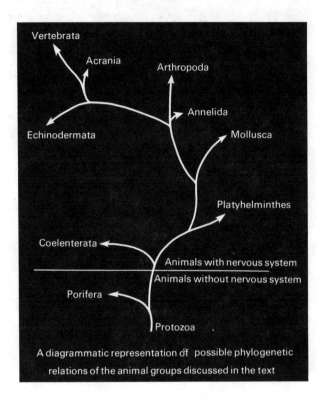

Vertebrata

Acrania

Arthropoda

Annelida

Echinodermata

Mollusca

Platyhelminthes

Coelenterata

Animals with nervous system

Animals without nervous system

Porifera

Protozoa

A diagrammatic representation of possible phylogenetic relations of the animal groups discussed in the text

The complexity of control systems in the various invertebrate phyla varies greatly, ranging from an absence of nervous and endocrine systems in the protozoans to the well-developed systems found in the insects and octopuses. As we have already dealt with unicellular protozoans, we shall arrange the multicellular invertebrate phyla with respect to the complexity of their nervous systems. Whether this indicates the path along which the nervous system has evolved is a matter of conjecture, because the relationships between many phyla are not clear. Zoologists are currently debating such questions as 'Were the molluscs derived from an annelid-like ancestor or from a platyhelminth-like one?'.

65

COELENTERATA (SEA ANEMONES, JELLYFISH, CORALS, ETC.)

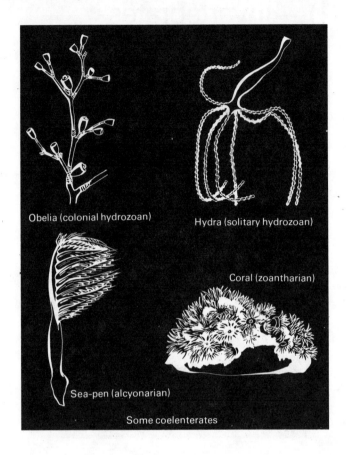

Obelia (colonial hydrozoan)

Hydra (solitary hydrozoan)

Coral (zoantharian)

Sea-pen (alcyonarian)

Some coelenterates

These animals are all aquatic; most of them live in the sea, though there are a few freshwater species. The phylum name means 'hollow gut', and refers to the fact that the main body cavity is the digestive cavity. Stinging cells, usually carried on tentacles, are used for prey capture and defence. The group is of particular interest because it is the most simply organized of those which possess a nervous system. Thus we might expect the system to be relatively simple, and perhaps the coelenterate condition represents the basic plan from which the more complex systems evolved.

The nervous system is diffuse and lacks definite centralization. A *nerve network* of bipolar, tripolar and multipolar neurons (Fig. 41) occurs beneath the outer layer of cells (*epidermis*) and also beneath the cells lining the

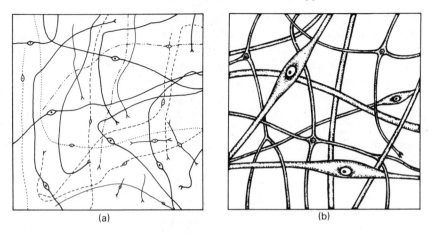

(a)   (b)

FIG. 41: Nerve networks in coelenterates: (a) portion of the nerve network in the mesentery of the sea anemone *Metridium senile* (AFTER BATHAM, PANTIN AND ROBSON), and (b) epidermal nerve network from a jellyfish (FROM BULLOCK AND HORRIDGE)

digestive cavity. The synapses between neurons are mainly *unpolarized*; transmission of impulses can occur in either direction. Conduction of impulses over appreciable distances in a nerve network is slow because many neurons and many synapses are involved. However, conduction can occur around a cut in the network; the nervous pathway then involves neurons still intact at the sides of the cut. When a train of impulses arrives at a synapse, *summation* may occur (p. 40) and, in many but not all synapses, another property is present, that of *facilitation*. Repetitive stimulation is necessary to demonstrate facilitation, where the strength (*amplitude*) of the response depends on the *frequency* of stimulation. Thus, in anemones, electrical stimulation of the body wall shows that contraction depends not on the strength of the stimulus, but on the frequency and number of stimuli. A single stimulus does not evoke a response, a second stimulus soon after the first results in a slight contraction, and repetitive stimuli lead to increasing increments of contraction until a limit is reached.

The different muscles of anemones (Fig. 42) are excited by different frequencies of stimulation, some requiring a much higher frequency than others before they contract. The closure of an anemone, involving shortening of the column, withdrawal of the tentacles and final constriction of the rim of the column to enclose the tentacles, is permitted by the responses of its various muscles, which are organized in such a way that they operate in sequence. Betty J. Batham and C. F. A. Pantin found that in the anemone *Metridium* electrical stimuli applied to the column wall at a frequency of 1 per 10 sec. resulted in contraction of the circular muscles of the column

67

only. The parietal muscles which shorten the column were excited at a frequency of 1 per 3 sec. A frequency of 1 per 2 sec. gave greater parietal contraction, and in addition caused a response of the longitudinal retractors in the mesenteries which withdraw the tentacles. One impulse per 1.5 sec. gave rapid retractor responses and a slow response by the sphincters which constrict the column rim. At a frequency of 1 per 0.6 sec. the sphincter muscle operated rapidly. Thus we can see that the muscle which operates last in this sequence, the sphincter, requires the highest frequency of stimulation, and this ensures that the column ring is not constricted before the tentacles have been withdrawn.

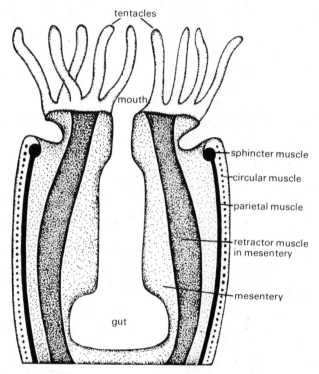

FIG. 42: *Metridium* in vertical section to show musculature (AFTER BATHAM AND PANTIN)

Anemones are capable of performing the closure reflex relatively quickly, and this involves a *through-conducting system*. This is a network of neurons with elongated processes located around the mouth region (oral cone) and longitudinally in the mesenteries which contain the longitudinal retractor muscles. In *Calliactis*, the rate of conduction in such a system is rapid,

1.0-1.2 m/sec. compared with a rate of 0.1 m/sec. vertically in the column wall. It is thought that the synapses in the through-conducting system cause no, or very little, delay in the passage of impulses, and perhaps these are electrically transmitting in a fashion similar to those of some crayfish and earthworms.

In the jellyfishes, the coelenterate nervous system reaches its highest level of differentiation, with the development in many species of nerve rings around the rim of the animal or sometimes marginal swellings (ganglia). Perhaps here we can see the beginnings of the development of a CNS. The first development of *true sense organs* appears in the jellyfish. These are usually confined to the rim and may be sensitive to light, gravity or chemicals.

A considerable repertoire of behaviour patterns is shown by coelenterates and, despite the relative simplicity and often diffuse organization of the nervous system, these reactions are well integrated. Some behaviour patterns, such as rhythmical contractions of the column muscles and feeding in the anemone *Metridium*, apparently can arise spontaneously—in the absence of any stimulus from the environment.

The freshwater species *Hydra* will stop responding to mechanical shaking and remain fully expanded if it is exposed to repeated mechanical stimulation every sixteen seconds for about three hours. Anemones often contract when a drip of water falls into their aquarium, but after very few repetitions, contraction ceases. Some people claim that coelenterates are capable of more complex types of learning, e.g. learning by association. Donald Ross has described how the swimming anemone *Stomphia* will detach from its position and swim away if touched by *Dermasterias*, a starfish predator. After twenty trials, pairing mechanical pressure to the base of the column (which causes closure of the anemone) with touch by the starfish, the normal reaction of escape was not shown when touched by the starfish alone.

EVOLUTION OF THE COELENTERATE NERVOUS SYSTEM

It is possible that diffuse organization and the lack of extensive centralization in the coelenterate nervous system are related more to the radially symmetrical body form and sessile habit than to evolutionary antiquity of the nervous system. However, many zoologists believe that the 'coelenterate type' of nervous system is the one from which all others have evolved. Basing considerations on the types of organization to be found in the coelenterates, a number of theories for the origin of the nervous system have been proposed. For many years G. H. Parker's theory has been accepted by most zoologists. Parker suggested that 'in the beginning' there were only independent effectors, structures which could respond to stimuli without innervation, e.g. contractile cells (*myocytes*) of sponges and the stinging cells of coelenterates. The second stage visualized was the development of receptor cells from undifferentiated epithelial cells. These were located adjacent to the muscle cells and affected them. The final step involved the incorporation of 'protoneurons' between the receptor and effector. All of these 'stages' in evolution can be seen in the condition of the nervous system in present-day coelenterates.

Parker's theory has been criticized by a number of recent workers who point out that a 'triad' of receptor, conductor and effector working in isolation would confer no advantage whatsoever on its possessor. 'Integration' is lacking in such a condition, and this function is as fundamental as conduction in any nervous system. A system where the inputs from numbers of receptors are co-ordinated is necessary for integration to occur.

A more recent theory by L. M. Passano proposes that individual primitive muscle cells first evolved into groups and this permitted more extensive movements of the body. Some of these cells became *pacemakers*, their activity arising spontaneously and involving a membrane depolarization which is envisaged as spreading passively to adjacent cells, thus influencing their contractile activities. Recurrent feeding movements could thus be performed by groups of muscle cells. At this stage, changes in membrane potential involved only passive depolarization spread both in the muscle cells and in the pacemakers which are destined to evolve into neurons. Examples of localized areas, each having its own pacemaker regions with regional autonomy, occur in many coelenterates. Later, with the evolution of an all-or-none response by the neurons, a nerve network arose and the various autonomous regions were linked. Some pacemakers could then become specialized for a general control of the whole organism, while others became subordinate centres controlling specific activities. Through-conducting strands and nerve rings could now develop, and also localized concentrations of neurons (ganglia). Receptors could be grouped into sense organs in association with these ganglia. Examples of each stage visualized in this theory can be obtained from within the phylum Coelenterata.

PLATYHELMINTHES (FLATWORMS)

These may be free-living or parasitic, but the parasitic forms, such as the flukes and tapeworms, have a poorly developed nervous system and will not be considered here. The free-living forms (turbellarians) are aquatic, or if terrestrial, live in moist surroundings. The platyhelminths are the 'lowest' group of the bilateral phyla to possess a centralized nervous system. A distinct brain is present from which a number of longitudinal cords arise, and these are linked by *commissures*. The basic plan of the CNS

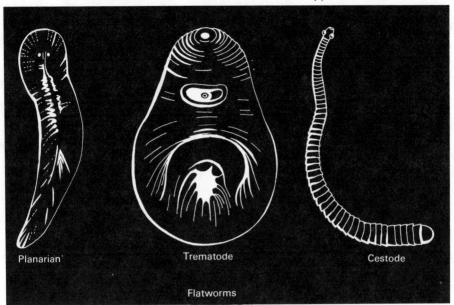

Planarian   Trematode   Cestode

Flatworms

is a brain plus from three to five pairs of longitudinal cords, with fairly regular commissures (Fig. 43).

Some platyhelminths do not possess obvious longitudinal cords in their nervous system, but an undifferentiated *plexus*. This condition is thought to be derived from the basic condition, in a similar fashion to the nervous system of triclads (planarians) which often have two ventral longitudinal cords.

The brain may be absent in some platyhelminths which have very simple nervous systems, and whether this condition is primitive or not is a matter of dispute. When the brain is present, as in most species, it ranges in structure from a simple node of nerve fibres to ganglia with numbers of different types of neurons enclosed. Here differentiation into an outer *rind* of cells and an inner *core* of fibres occurs, and *tracts* appear between cell groups. Within the brain and possibly in the cords, unipolar neurons are common—this condition shows an advance over that of coelenterates.

It is possible that the cords of the CNS are homologous throughout all the bilateral phyla. One theory, the *orthogon theory*, suggests that primitively there were about six or eight longitudinal concentrations of nervous tissue, as seen in some platyhelminths and also in ctenophores (comb jellies). These cords were linked by transverse commissures. By loss of some cords, emphasis of others, or by fusion in the midline of the dorsal cords, we can obtain the conditions seen in annelids, arthropods or vertebrates (Fig. 44).

The *peripheral nervous system* of platyhelminths consists mainly of a

71

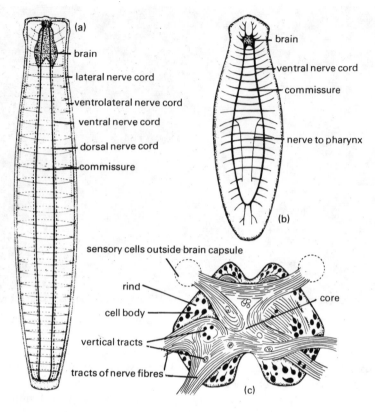

FIG. 43: The nervous system in flatworms: (a) nervous system of a primitive platyhelminth (*Alloeocoela*) (AFTER REISINGER), (b) nervous system of the triclad *Procerodes* with the reduced number of longitudinal cords (FROM HYMAN), and (c) section through a polyclad brain (AFTER HADENFELDT)

submuscular plexus which is continuous with the main longitudinal cords and commissures. A plexus is particularly well developed in the pharynx and can control the activities of the pharynx in isolation, even allowing it to feed. The pharynx is normally inhibited from its spontaneity by its connection with the ventral cords.

Receptor organs in platyhelminths range from single sense cells to multicellular complex eyes. The single sense cells may be tactile, chemoreceptive or rheoreceptive (sensitive to water currents) in function. Multicellular sense organs may include patches of sensory epithelium, papillae and depressions, and also eyes and gravity receptors.

Removal of the brain has severe effects in some platyhelminths, e.g. polyclads, which often suffer cessation of locomotory activities and

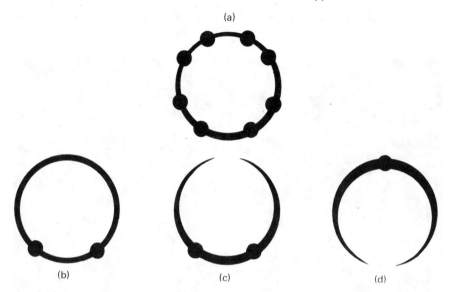

FIG. 44: The orthogon theory (AFTER HANSTRÖM): diagrammatic sections through the central nervous system of (a) a primitive platyhelminth showing four pairs of longitudinal cords (see also Fig. 43), (b) an annelid, (c) an arthropod, and (d) a vertebrate

inability to find food. In triclads, brain removal is much less serious though locomotion may be slower than normal and difficulty is encountered in locating food.

Planaria appear capable of learning, and even complex types of learning have been claimed. They will habituate to rotation of the dish which contains them. The initial response is a brief cessation of the gliding movement, but with repetition of the stimulus at one-second intervals the response fails after about twelve trials. Such cessation of a particular response as a result of experience also occurs when planaria are repeatedly stimulated by light or mechanical vibration.

Classical conditioning has been demonstrated in planaria. When light is presented with electric shock for about 150 trials, planaria eventually turn or shorten their body when given light alone. Before the training period, light alone resulted in no significant response by the planaria. The experimental approach in this work has been criticized by a number of workers, but there is now a sufficient amount of evidence to give a more or less definite 'yes' to the question 'Can planaria learn?'. Cognitive memory of their environment has also been claimed for some planaria which show an increased delay in the time taken to feed when introduced into a new versus a 'familiar' environment.

ANNELIDA (SEGMENTED WORMS)

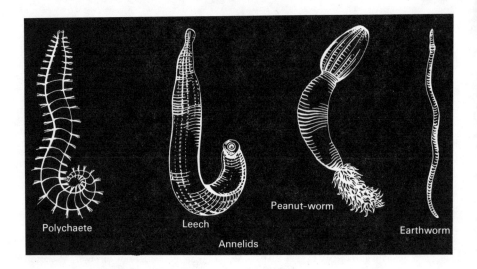

Polychaete   Leech   Peanut-worm   Earthworm

Annelids

The earthworms and numerous marine worms, such as the ragworms and tubeworms, belong to this phylum. The special advance in body structure shown by the annelids is the development of segmentation. In each segment certain structures are repeated, such as nerve ganglia, appendages and excretory organs. Anteriorly a definite head with numerous sensory structures is developed.

The CNS of animals first developed directly below the superficial epidermal layer (*basi-epithelial* in position), and this primitive condition still exists in a few annelid species. This is interesting because most other annelids have their CNS in a completely internal position, free in the body cavity. Even the platyhelminths have the CNS positioned internally.

Basically the CNS of annelids consists of a pair of cerebral ganglia (the brain) situated above the oesophagus, with connectives passing around the gut to suboesophageal ganglia below. From the suboesophageal ganglia two ventral longitudinal cords pass along the length of the animal, with ganglia in every segment. The ganglia of each segment are linked by commissures, forming a ladder-like structure (Fig. 45a), but usually the two ventral cords are fused into one mid-ventral cord (Fig. 45b).

A subepidermal plexus is retained in annelids, but this is mainly associated with epidermal sense organs. However, a few connections are made with muscle cells and also with the CNS. A visceral nervous system is present and consists of a number of neurons forming a plexus on the anterior part of the

74

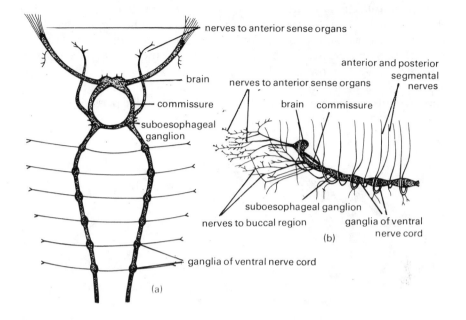

FIG. 45: The nervous system in annelids: (a) brain and anterior region of the ventral nerve cords in the tubeworm *Serpula* (FROM SEDGWICK), and (b) lateral view of brain and anterior region of the ventral nerve cord in the earthworm *Lumbricus* (FROM BULLOCK AND HORRIDGE)

alimentary canal. Connection is made with both the cerebral and sub-oesophageal ganglia.

Within the ganglia of the CNS the structural arrangement follows the typical invertebrate plan of cell bodies on the outside (the rind) and fibres disposed centrally (the core). Nerves are given off from each segmental ganglion; two, three or four pairs is the common condition. The supra-oesophageal ganglia give off many nerves to sensory structures such as tentacles and eyes.

Within the ventral cords of annelids *giant nerve fibres* may occur. These vary in structure and may be multi- or unicellular, septate or non-septate, and it is thought that they have evolved separately in the various animal groups which possess them. These large-diameter nerve fibres can transmit impulses at much faster rates than normal-sized fibres and are associated with the performance of 'startle' or 'escape' reactions. Escape reactions require that many muscles should be excited as symmetrically and synchro-nously as possible; thus rapid impulse conduction rates are vital. The earthworm *Lumbricus* has three giant nerve fibres within the ventral nerve

75

cord (Fig. 46). The two lateral giants have a smaller diameter (about 50 µ) than the median giant (about 75 µ). Mechanical stimulation at the anterior end of the worm excites the median giant fibre, which conducts at a rate of from 15 to 45 m/sec. Setae at the posterior end of the worm effect a grip on the sides of its burrow and the longitudinal body wall muscles contract rapidly, resulting in swift retreat to a safer position inside the burrow. The lateral giants respond to stimuli applied to the posterior parts of the worm.

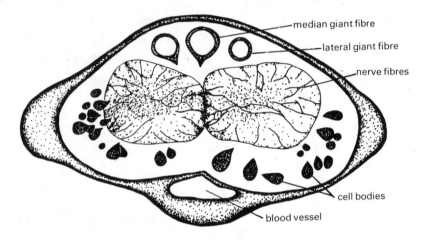

FIG. 46: Transverse section of the ventral nerve cord of *Lumbricus*
(AFTER BULLOCK AND HORRIDGE)

Due to the method of locomotion of many annelids, involving alternate contractions of the body and extensions in length, the nervous system is subjected to stretching. In fact, the stretch is often such that the nerve cords are reduced in diameter. Despite the reduction in neuron diameter the velocity of impulse conduction is not decreased.

Removal of the cerebral ganglia does not prevent the performance of certain activities such as locomotion, righting and maze-learning, while feeding and burrowing are affected to a degree. The treated worms show increased excitability and restlessness. Removal of the suboesophageal ganglia reduces muscle tension, search movements and spontaneity.

The brains of oligochaetes (earthworms, etc.), hirudineans (leeches) and some polychaetes (marine worms) are quite simple, and organs of special sense are often poorly represented. However, other polychaetes have exceedingly complex brains, comprising three distinct regions: the fore-, mid- and hindbrain. In such polychaetes optic ganglia are found for the first time. Within the brains of many polychaetes are mushroom-shaped

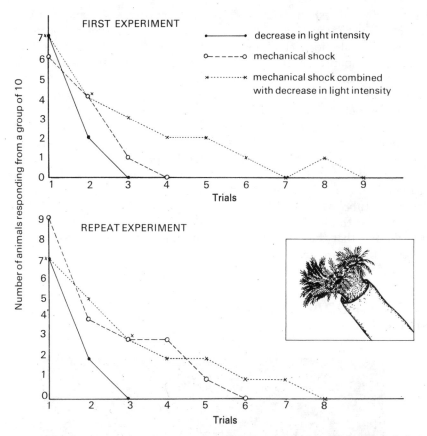

FIG. 47: Results of two experiments to demonstrate habituation in the tubeworm *Galeolaria*: withdrawal occurs in response to decrease in light intensity and to mechanical shock, but when either of these stimuli is given repeatedly the animal stops responding after a few trials. When the two different stimuli are presented together the worms take longer to habituate than to the single repeated stimulus. The time interval between trials was from 15 to 60 seconds depending on time taken for the animal to re-emerge from the tube. (AFTER THORNE)

structures called the *corpora pedunculata*. These are probably homologous with the regions of the same name in arthropod brains, where they are regarded as the highest association centres in that phylum.

Many worms live in burrows or tubes and are thus not exposed to the variety of stimuli that may affect the free-ranging and more active worms. The behavioural repertoire of these sedentary types might be expected to be more limited. A common behaviour pattern is a fast withdrawal response shown to such stimuli as a passing shadow, mechanical shock and touch.

The withdrawal response ceases after a few repetitions of a simple stimulus (e.g. passing shadow), but when the stimulus is more complex (e.g. shadow and mechanical shock combined) many more stimuli must be presented before the response disappears (Fig. 47). Some tubeworms have the ability to select suitable materials for tube construction, some choosing from a wide variety of materials, others selecting only flat pieces of shell or sand.

A number of earthworms and marine ragworms (e.g. *Nereis*) are capable of learning to make a choice of arms in a simple T-maze. If *Nereis* is positioned at the entrance of a glass tube it will crawl into it and through it—and will continue to do so for an extended period. If the worm is given an electric shock at the end of the tube, the time taken to crawl along it is increased until finally the worm will refuse to enter. Using 'refusal to crawl down the tube' as a criterion of learning, R. B. Clark found that *Nereis* specimens with the brain removed learned almost as quickly as intact animals, but that the retention time of brainless worms was considerably less than that of intact ones.

ARTHROPODA (INSECTS, CRUSTACEANS, SPIDERS, ETC.)

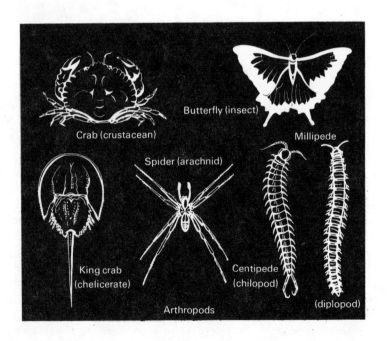

Arthropods

Crab (crustacean)

Butterfly (insect)

Spider (arachnid)

Millipede

King crab (chelicerate)

Centipede (chilopod)

(diplopod)

The basic arrangement of the arthropod nervous system is similar to that of annelids. The dorsal brain is linked to the suboesophageal ganglion by connectives around the gut. The ventral cord shows a segmental pattern and in primitive conditions contains a pair of ganglia in each segment, joined longitudinally by connectives and transversely by commissures (Fig. 48a). These commissures are usually lost due to the lateral fusion of the ganglia in each segment (Fig. 48c, d and e). The number of ganglia in the ventral cord varies greatly; some millipedes have about 200 while some insects have only one, as in the water measurer *Hydrometra*. The house fly *Musca* has a suboesophageal ganglion and one thoracic ganglion (Fig. 48e).

The first ganglion of the ventral cord, the suboesophageal ganglion, is the result of fusion between ganglia from two or more segments. The remaining ganglia are often condensed to a greater or lesser degree; less highly organized arthropods show little or no condensation (Fig. 48a, b and d) while the more highly organized forms may show extreme condensation (Fig. 48e). Longitudinal fusion between the connectives of the cord may occur, at least for part of their length (Fig. 48c). Giant fibres are common in the ventral cord.

The visceral nervous system, which supplies the anterior region of the gut, arises from the brain or oesophageal commissures and connectives, and consists of pairs of nerves, often with associated ganglia. The hindgut is supplied from the last abdominal ganglion.

There are three main regions to the brain: the fore-, mid- and hindbrain (Fig. 48f). The forebrain houses the optic ganglia and also the corpora pedunculata (in Crustacea the corpora pedunculata occur in the eyestalks)— thought to be centres for 'higher levels' of integration. Electrical stimulation of regions of the corpora pedunculata in bees can cause a variety of activities such as cleaning and aggression.

The phylum Arthropoda is the largest and probably the most diverse phylum in the animal kingdom, and the behaviour of its members varies greatly. Some species almost appear to be stimulus-bound and machine-like in their actions, for example fly larvae and woodlice reacting in a predictable way to such stimuli as light and humidity. Others (e.g. bees) have the ability to communicate, to learn spatial relationships, and to time their activities accurately.

Examples of all the major 'types' of learning have been described for arthropods, with the exception of 'insight' learning. Even the thoracic ganglion of a headless cockroach shows evidence of learning. In an experimental situation Adrian Horridge suspended a cockroach, and whenever one leg fell below a particular position an electric shock was given. Eventually the leg was held permanently above this critical position.

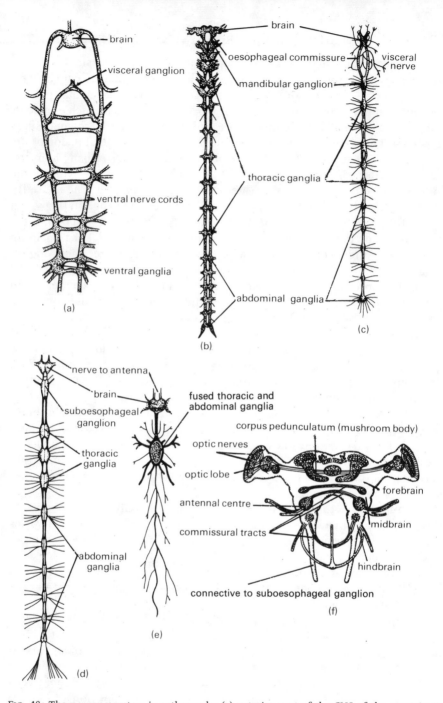

FIG. 48: The nervous system in arthropods: (a) anterior part of the CNS of the crustacean *Apus*, showing the condition in which the ventral cords and ganglia are well separated (FROM BARNES), (b) the CNS of the crustacean *Apseudes*, showing the condition in which the segmental ganglia are fused while the ventral cords remain separate (FROM SEDGWICK), (c) the CNS of the lobster *Homarus*, showing the condition in which the segmental ganglia and ventral cords of the abdomen are fused (FROM BULLOCK AND HORRIDGE), (d) the CNS of the earwig *Forficula* (AFTER IMMS), (e) the CNS of the housefly *Musca*, showing the condition in which thoracic and abdominal ganglia are fused (FROM IMMS), and (f) the brain of an insect showing the major fibre tracts (AFTER SNODGRASS)

MOLLUSCA (CHITONS, SNAILS, SLUGS, CLAMS, COCKLES, OCTOPUSES, ETC.)

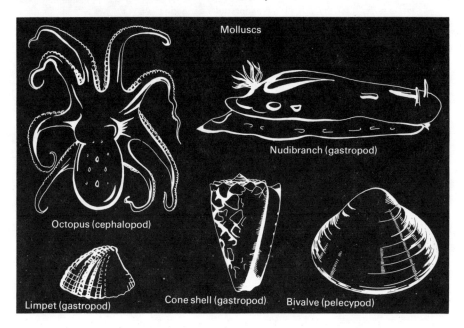

Molluscs

Nudibranch (gastropod)

Octopus (cephalopod)

Limpet (gastropod)   Cone shell (gastropod)   Bivalve (pelecypod)

The range of complexity in the nervous system of this group is extreme. In chitons the condition is less complex than in many flatworms. No distinct cerebral ganglia are present, the only ganglia being those connected with the buccal mass and its sense organs (Fig. 49). In contrast, the brain of cephalopods (squids and octopuses) is extremely complex, consisting of many fused ganglia (see Fig. 52), and rivals the condition seen in some vertebrates.

The basic arrangement of the molluscan nervous system includes about six pairs of ganglia linked by commissures and connectives (Fig. 50). Cerebral ganglia occur above the oesophagus and connectives pass around the gut to the *pedal ganglia* which control the activities of the foot. On the cerebro-pedal connectives, *pleural ganglia* occur. From these, longitudinal connectives pass back to the *intestinal* and *visceral ganglia. Buccal ganglia* innervate the buccal mass. This is the condition seen in the gastropods (snails, slugs, etc.).

The bivalves (clams, cockles, oysters, etc.) have a simplified nervous system (Fig. 51), this being correlated with their more or less sedentary mode of life. A distinct head with sense organs is lacking in the bivalves, the major sense organs occurring on a flap of tissue (the mantle) directly

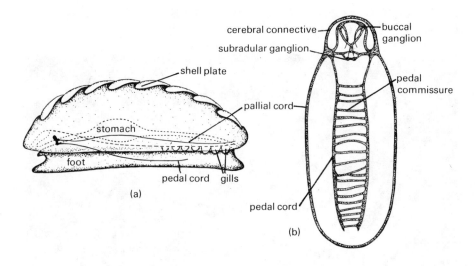

FIG. 49: The nervous system in chitons: (a) the position of the CNS in a chiton, and (b) the CNS of *Acanthochiton discrepans* (FROM HYMAN)

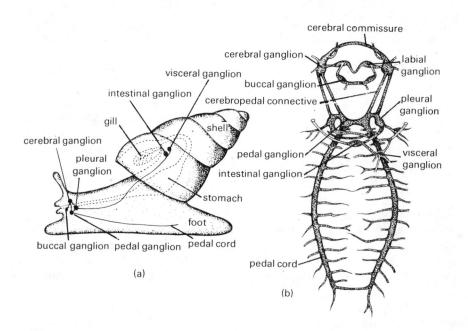

FIG. 50: The nervous system in gastropods: (a) the position of the CNS in a gastropod, and (b) the CNS of the limpet *Patella* (FROM HYMAN)

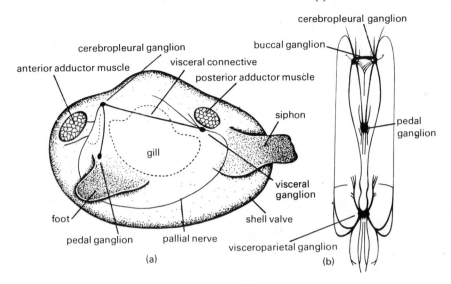

FIG. 51: The nervous system in bivalves: (a) the position of the CNS in a bivalve, and (b) the CNS of the bivalve *Sphaerium* (FROM BULLOCK AND HORRIDGE)

beneath each shell valve. In the cephalopods, however, the differentiation of the brain is on a scale unparalleled by other invertebrates. About thirty separate lobes occur in the octopus brain (Fig. 52), each of which has a characteristic structure. The learning ability of octopuses surpasses all other invertebrates; they are capable of making a wide variety of visual, tactile and chemotactile discriminations. They can recognize many different geometric shapes and their orientation; the results from numerous experiments involving different figures indicate that octopuses tend to classify objects on the basis of their horizontal and vertical extents (see p. 10), but they also utilize additional bases for classifying objects. They can recognize the size of objects seen, and also the range at which an object occurs.

Blinded octopuses can make tactile discriminations using information from sense organs in the suckers. Martin Wells found that objects of differing texture could be distinguished, but no discrimination was made between two similar objects which were different in weight. Wells suggests that proprioceptive information—information about the positions or movements of the parts of its body—does not penetrate to those regions of the brain concerned with learning. It has been found, mainly using ablation techniques where regions of the brain are removed, that the highest regions (topographically) of the brain are those concerned with memory, and that there are two almost independent learning systems, one visual, the other chemotactile, and these are situated in different brain lobes.

83

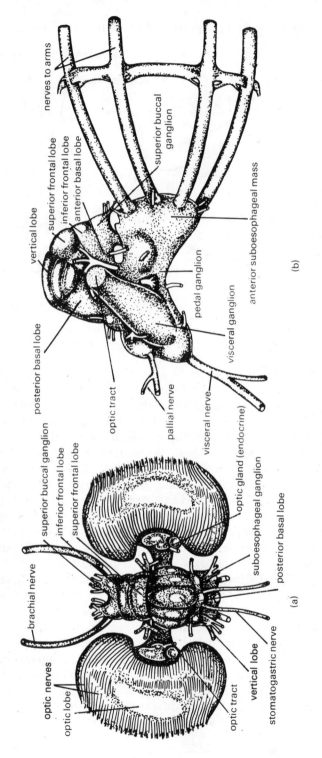

nerves to arms

superior buccal ganglion

superior frontal lobe
inferior frontal lobe
anterior basal lobe

vertical lobe

anterior suboesophageal mass

pedal ganglion

visceral ganglion

posterior basal lobe

optic tract

pallial nerve

visceral nerve

optic gland (endocrine)

(b)

brachial nerve

superior buccal ganglion
inferior frontal lobe
superior frontal lobe

optic nerves

optic lobe

suboesophageal ganglion

posterior basal lobe

vertical lobe

optic tract

stomatogastric nerve

(a)

FIG. 52: The nervous system in cephalopods (AFTER YOUNG): (a) dorsal view of octopus brain,
and (b) side view of octopus brain with optic lobe removed

ECHINODERMATA (STARFISH, SEA URCHINS, ETC.)

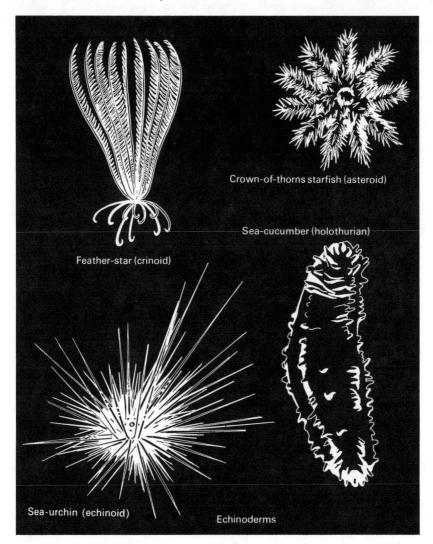

Crown-of-thorns starfish (asteroid)

Sea-cucumber (holothurian)

Feather-star (crinoid)

Sea-urchin (echinoid)

Echinoderms

These are exclusively a marine group and have a *pentaradial symmetry* (most obvious perhaps in many starfish with five arms). This symmetry and the lack of a distinct anterior end have important consequences on the internal structures of the animals, including the nervous system. Some similarities to the coelenterate condition are shown in the radial organization of the major nervous pathways, the presence of extensive nerve plexuses and the occurrence of small-sized neurons.

85

Superficial and deep nerve plexuses occur and in places these are concentrated into nerve cords, often in a ring around the oesophagus and along each radius. In the starfishes (asteroids), the superficial plexus is mainly sensory and located just beneath the epidermis (Fig. 53). The plexus is condensed into three definite nerve tracts: (1) the circum-oral nerve ring around the oesophagus, (2) the radial nerve of each arm, and (3) the adradial nerve cords, two in each arm lateral to the centrally positioned radial nerve. The deeper plexus is mainly motor, and from it various tracts arise and pass to the muscles.

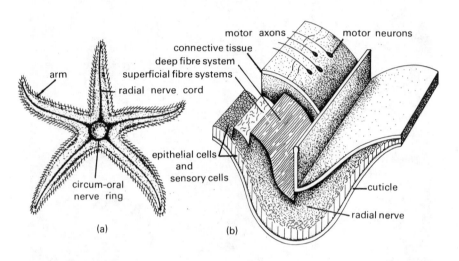

FIG. 53: The nervous system in echinoderms: (a) diagram showing the position of the CNS in a starfish, and (b) stereogram of a portion of the radial nerve of a starfish (AFTER J. E. SMITH)

An isolated starfish arm is capable of co-ordinated movements, but for the normal movement of the whole animal the oral ring must be intact. During directed movement one arm is dominant and leads; with change of direction there is a transfer of dominance. The nerve centres involved in such activities are at the junction of the radial nerves with the oral ring.

Behaviourally, some of the echinoderms have not progressed very far. The sea cucumbers (holothurians) spend a very sluggish existence lying on or burrowing in the bottom sand and mud, eating whatever happens to present itself. Their response to prodding is often negligible, but if they are roughly handled, tubules may be shot out of the rectum. With still rougher treatment the animal may eviscerate. Starfish have a larger repertoire of responses. They move more rapidly and can right themselves in a number of ways if they are turned over. They have a more complex feeding behaviour, often attacking bivalves by opening the shell valves slightly and everting their stomach to digest the tissues *in situ.*

TRENDS IN THE ELABORATION OF THE NERVOUS SYSTEM IN THE INVERTEBRATES

We have seen that the nervous system of coelenterates, considered primitive, consists of networks of bipolar and multipolar neurons. Special advances made by some members of the group are the increased length of the neuron processes in the anemone 'through-conducting system', and the condensation of neurons into a ring and/or ganglia in some jellyfish.

Tendencies towards specialization of the nervous system in the remaining bilateral phyla include:

CONDENSATION

Many neural elements become concentrated into the ganglia and cords of the CNS, as distinct from the peripheral nervous system. The most common number of longitudinal cords is two (in annelids, arthropods and some platyhelminths), possibly derived from an ancestor with six or eight cords (orthogon theory).

CEPHALIZATION

A head is developed at the anterior end of the bilateral animal. This region becomes well supplied with sense organs receiving innervation from the brain which, in the more highly organized animals, becomes extremely complex and vital for their continued existence. Cephalization is extreme in such forms as the octopus.

DEVELOPMENT OF GANGLIA

The cords of the CNS arose by a condensation of a network or plexus of bipolar and multipolar neurons. Primitively, the neuron cell bodies were scattered along the length of the cords. This condition can still be seen in chitons and some platyhelminths. With elaboration of the cords, the neuron cell bodies become localized into ganglia so that the cords consist of fibres only. In addition, the primitive subepithelial position is lost and the cords come to lie in the main body cavity (annelids provide examples of all grades of development from a superficial, subepidermal position of the cords to a completely internal position).

THE BEHAVIOUR OF ANIMALS

DIFFERENTIATION

The platyhelminths have developed a brain and even in these simple animals the basic organization of the invertebrate CNS is evident: a rind of cell bodies and a core of fibres and synapses. In both brain and cords, further differentiation occurs and the core comes to contain tracts (of fibres only) and neuropile tissue (a plexus of fibres with dendrites, axon endings and synapses). The neuropile can be considered to be a region for integration. Mushroom bodies (corpora pedunculata), which are the highest integration centres of the brain, appear first in some polychaetes and are prominent in many arthropods.

DEVELOPMENT OF GIANT NERVE FIBRES

A system of giant fibres is developed within the CNS of those animals showing a fast escape reaction or startle response. The fibres may be formed by the enlargement of a single axon or by the fusion of many axons. Selection pressure favouring the development of larger diameter giant fibres which would allow even faster escape responses is countered, at least in part, by the limitations of space within the organism and within the nerve cords themselves. The space occupied by a single giant fibre could be filled with hundreds of small-diameter fibres giving a significantly greater information-carrying capacity to the animal.

EVOLUTION OF HORMONES

Hormonal control of behaviour in invertebrates is little understood (see p. 56), but we know even less about the origin of hormones. The first hormones to appear in the animal kingdom are thought to have been the products of nerve cells or pre-nervous elements. The first stage was probably an immediate utilization of the secretion from a group of cells, as in the present condition of flatworms. Here a group of neurosecretory cells in the brain produce a hormone influencing regeneration. Later, a storage region evolved for the neurosecretion (a *neurohaemal organ*). The secretion (Fig. 54) passed from the site of production in the nerve cells, along a tract composed of nerve cell processes, to the storage site (see also Fig. 36, p. 58). Large amounts of hormone could now be released as required. Neurohaemal organs are seen in such animal groups as the annelids, crustaceans, insects and vertebrates. In insects the *corpus cardiacum* is such an organ, corresponding to the *neurohypophysis* of vertebrates which receives its hormones via tracts arising from the hypothalamus of the brain.

Subsequent to the development of neurohaemal organs, and with the increasing complexity and organization of tissues, came the formation of

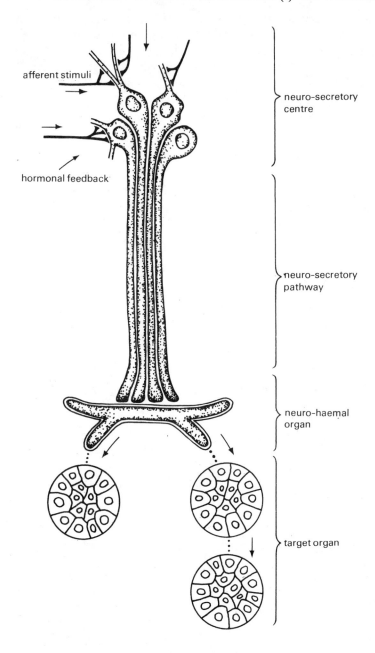

afferent stimuli

hormonal feedback

neuro-secretory centre

neuro-secretory pathway

neuro-haemal organ

target organ

FIG. 54: Neurosecretory system of animals (AFTER SCHARRER AND SCHARRER)

non-nervous endocrine glands in various regions of the body in both vertebrate and invertebrate animals. Examples of such glands are the *prothoracic glands* of insects and the thyroid, pancreas and gonads of vertebrates. Even though non-nervous endocrine glands assume very important functions, the overall control of the endocrine system is effected by neurosecretory cells of the central nervous system.

5 Evolution of Control Systems (II)—Vertebrates

Vertebrate animals exhibit a great diversity of structure and physiology, but the discontinuity between the major classes of vertebrates is less marked than that in many invertebrate groups. For this reason the comparative approach in anatomy is most fruitful in clarifying evolutionary relations among the vertebrates. We shall take this approach to study the evolution of various structures in vertebrate control systems.

The general patterns of the vertebrate body may be illustrated by the following features. The first two are shared by many higher invertebrates, while the rest are unique to the vertebrates.

GENERAL PATTERNS OF THE VERTEBRATE BODY

- *Segmentation:* The basic structure shows the development of various organs, such as muscles and nerves, in each segment of the body.
- *Bilateral symmetry:* Apart from the digestive system the basic structure is bilaterally symmetrical.
- *Notochord:* This axial, elastic, rod-like structure is the major supporting organ of the body in some primitive forms, but is replaced by the vertebral column in higher forms.
- *Branchial clefts* (gill-slits): These are perforations of the pharynx which serve as respiratory openings for gills in lower forms. In higher vertebrates with lungs, these appear only in the embryonic stage and are later modified to serve endocrine and other functions.
- *Dorsal nerve cord:* This runs parallel and superior to the notochord or the vertebral column and is filled with fluid. This is the central nervous system of all vertebrates and its anterior portion is differentiated into the brain.
- *Appendages:* The higher forms have two pairs of appendages (fins, flippers or limbs). Each limb consists of tissues derived from several segments unlike the invertebrate appendages. (Each leg of a centipede, for example, is formed from one segment only.)
- *Tail:* If present, this is always post-anal in position, unlike segmented invertebrates.

The most important evolutionary feature is the notochord (*chorda*

91

dorsalis). Because the notochord appeared in evolution before the vertebral column, there are some animals which have a notochord but no vertebral column. On the other hand, all animals with a vertebral column have a notochord at some stage of their development. We group all those animals that have a notochord at some stage of their lives and call them *chordates* (phylum Chordata). In most chordates the notochord is replaced as the mechanical skeleton by a cartilaginous or bony vertebral column. Such animals belong to the subphylum Craniata or Vertebrata, whereas those without the vertebral column belong to the subphylum Acrania.

ACRANIA

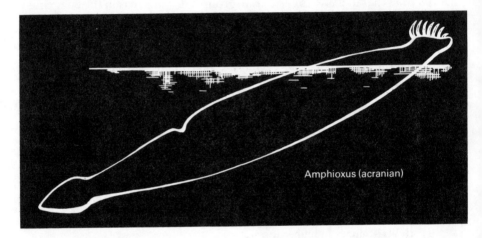

Amphioxus (acranian)

These animals are all aquatic and many are quite unlike vertebrates, as they have the free-living stage only during larval life. The cunjevoi (sea-squirts), for example, have a sedentary adult life firmly attached to rocks in the lower tidal zone of the sea shore. Apart from the gill-slits (many thousands are present) perforating the sides of the branchial sac, their anatomical features give no indication of their affinity with vertebrates. It is only as free-swimming larvae that they possess a rudimentary nerve cord with an eye-spot and a notochord.

However, another animal called the lancelet *(Amphioxus)* resembles vertebrates in many respects. It is a small animal about 5 cm in length, found in shallow seas with a sandy bottom. It is elongated and pointed at each end (hence named *Amphioxus*), with the body laterally compressed. This animal has no head, and the only indication of specialization of the nerve cord is a slight enlargement of the inner diameter of the hollow cord at its anterior end, called the *cerebral vesicle*. There are a number of pigment-spots in the nerve cord, and one of these located at the front end forms a primitive eye-spot.

The nerve cord is the central nervous system of the animal and runs above the notochord as a straight tube. In each body segment, one dorsal nerve root (*dorsal nerve*) and one bunch of ventral nerve roots (*ventral nerves*) arise from the nerve cord (Fig. 55). The ventral nerves of each

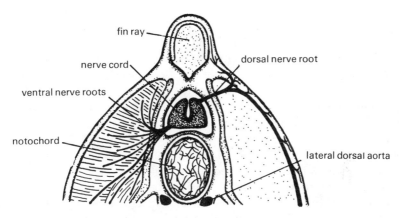

FIG. 55: The dorsal portion of a transverse section through the body of *Amphioxus* showing the dorsal and ventral nerves (AFTER GOODRICH). The corresponding nerves on the opposite sides do not normally appear in the same section.

segment supply only the muscle fibres of the body muscle (*myotome*) of that segment and are, therefore, *motor neurons* (efferent nerves). The dorsal nerves transmit nerve impulses received from sense organs and are, therefore, *sensory neurons* (afferent nerves). They also innervate the smooth muscles of the gut and the *atrium*, which is a special outer cavity of the animal protecting the pharyngeal region with the vulnerable gills. In *Amphioxus*, the information received by the sense organs is transmitted directly to the central nervous system as in many other invertebrates; it is only later in the evolution of the chordates that non-nervous sensory cells are found (see p. 45). In contrast to the vertebrates, no ganglia are found on the dorsal roots of the nerve cord. The two most anterior pairs of roots are dorsal (there are no ventral roots corresponding to them), and they innervate the sense organs of the snout and oral region.

Thus, the primitive conditions of the nervous system of *Amphioxus* include: very slight specialization of brain (*cerebral vesicle*); the derivation of all afferent nerve fibres from sensory cells; the absence of dorsal root ganglia.

LOWER VERTEBRATES

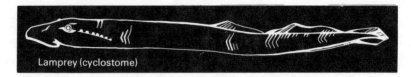

Lamprey (cyclostome)

The subphylum Craniata includes all those animals known as lampreys (Cyclostomata), fishes (Pisces), newts and frogs (Amphibia), reptiles (Reptilia), birds (Aves) and mammals (Mammalia). All possess a skull, a brain, a heart and red blood-corpuscles.

The lamprey *Petromyzon*, one of the lower vertebrates, is elongated like an eel and grows from about 10 cm to 1 m or more in length. It differs from all other vertebrates by the absence of jaws; there is a circular sucking mouth without any skeletal anchoring to the head skeleton. The lamprey has a definite head which is a specialization of the anterior region of the body, two types of nervous development being responsible for its formation:

- the development of paired sense organs;
- the correlated specialization of the nerve cord into a brain.

The organs of the head are protected by a special structure called the *skull*.

The olfactory nerves carry information from the epithelium (lining of cells forming the body's outer or inner layer) of the nasal sacs to the brain, and are concerned with chemical sense. The nasal sacs have a single median opening to the outside.

The paired eyes are first formed as the *optic vesicles*. In *Amphioxus* the cells sensitive to light line the nerve cord, but in vertebrates a portion of this wall protrudes sideways to form the optic vesicle on each side. The sensitive cells are still morphologically the inner side of the wall of the brain. The outer side of each vesicle is pushed in to form a cup (*optic cup*), and the lens which is developed from the superficial skin fits into the mouth of the cup. The cells on the inner layer of the cup are sensitive to light and form the *retina*. The outer layer contains the pigment. Outside this, there are two other layers: an inner layer, the *choroid*, contains blood vessels, and an outer layer, the *sclerotic*, is hard and has a protective function. The sclerotic is the outermost layer of the whole eyeball and in front of the lens it is transparent, forming the *cornea*. The cornea is in contact with the epidermis (skin layer), which is also transparent here, forming the *conjunctiva* (Fig. 56). In the lamprey, the two most anterior myotomes give rise to the *cornealis* muscle, the tendon of which is inserted into the rim of the conjunctiva. It pulls on the cornea from the side and flattens it. This in turn pushes the lens inward, closer to the retina. Thus, accommodation of vision is achieved by alteration of the shape of the eyeball. This is different from the methods

94

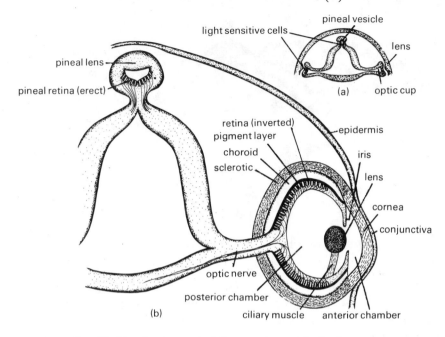

FIG. 56: The development of the eyes in vertebrates (AFTER DE BEER): (a) formation of the optic cup and the position of sensory cells, and (b) structure of pineal and paired eyes showing the relations of the retina

used by other vertebrates. The nerve fibres which convey the impulses away from the retina run between the layer of sensitive cells and the lens. Therefore, the image of the object perceived reaches these sensitive cells after passing through the layer of nerve fibres. The paired eyes of all vertebrates have this type of retina, called the *inverted retina*. The nerve fibres of the *pineal eye*, which develops as a pineal vesicle on the median line, run back to the brain without passing above the sensitive cells. This type of retina is called the *erect retina*. The pineal eye has a pigmented retina, a flat and imperfectly formed lens, and is capable of distinguishing differences in the intensity of light.

The ears of the lamprey do not serve the purpose of hearing, but are organs of balance. They take the form of sacs situated behind the eyes and give off canals which form half-loops, each opening into the sac. These are the *semicircular canals*, each one bearing a swelling (*ampulla*) containing a *statolith* (Fig. 57). The lamprey has two semicircular canals on each side. In all higher vertebrates there are three pairs of them in planes at right angles to one another. The semicircular canals are connected to a large

95

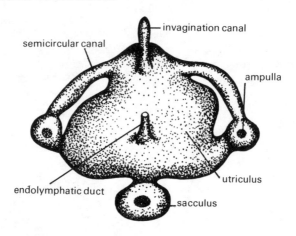

FIG. 57: Diagrammatic 'ear' of the lamprey *Petromyzon*—an organ
of balance (FROM PARKER AND HASWELL)

vestibule which consists of two chambers (the *utriculus* and the *sacculus*).
In these chambers, statolithic particles (*otoliths*) are supported on sensory
cilia. Gravity makes these particles press on the cilia, whatever the position
of the animal; therefore, according to which of the cilia are stimulated, the
animal is informed of its position relative to the vertical. The semicircular
canals contain fluid, the *endolymph*, as do all parts of the auditory sac. When
the animal starts or ceases moving, a flow of endolymph takes place in the
semicircular canals. The receptor cells in the ampullae detect this flow and
inform the animal as to its orientation and acceleration.

To facilitate development of these sense organs, the anterior region of the
nerve cord became modified and enlarged to form a brain. The brain (Fig.
58) can be divided into three major regions: forebrain, midbrain and hindbrain. The forebrain bears the *olfactory lobes* in front and optic nerves to the
side extending to the eyes. The roof bears two pineal eyes, lying in the
middle line, and the lower one (*parapineal*) is degenerate. Above the pineal
eye, the skull is thin and the tissues are more or less transparent. The
impulses from the photosensitive cells of the pineal eye are conveyed
through the *hypothalamus* to the *pituitary gland*. Secretion from a part of
the latter gland, as a result of stimulation by these impulses, is known to
cause the skin of the lamprey to darken. The hypothalamus is situated in
the ventral region of the forebrain, and the pituitary gland (*hypophysis*) is
closely applied to it. Between the hypophysis and the floor of the skull
there is a large pouch, called the *nasohypophysial sac* or *hypophysial sac*,
which ends blindly below the anterior end of the notochord and opens to
the outside through the nostril. This elongated pouch is compressible and

respiratory movements squeeze or relax its blind end, thus enabling the flow of water in and out of the pouch through the nostril for olfaction. The midbrain bears the imperfectly differentiated *optic lobes* and the hindbrain has a small rudimentary *cerebellum* on its roof. The roof of the brain in the lamprey is generally thin and membranous, but concentration of nerve cells occurs where the two spheres of the forebrain are connected (*transverse commissures*), in the optic lobes of the midbrain, and in the cerebellum.

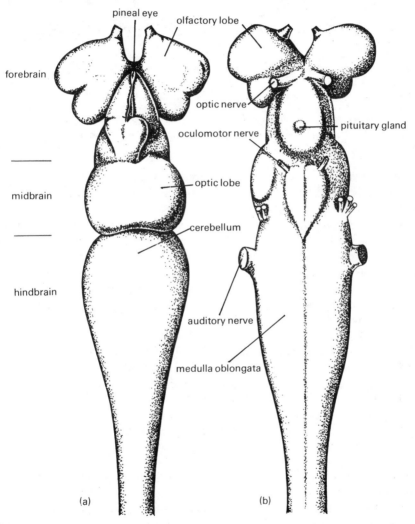

FIG. 58: (a) Dorsal and (b) ventral views of the brain of the lamprey (MODIFIED FROM PARKER AND HASWELL)

97

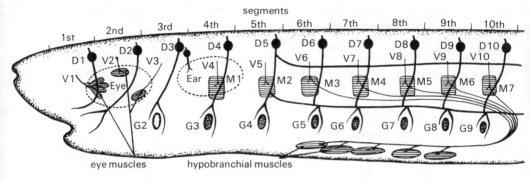

FIG. 59: Schematic diagram of the head and branchial regions (segments 1st to 10th) in the lamprey showing the segmental origin of dorsal nerves (D1 to D10), ventral nerves (V1 to V10), eye muscles, myotomes (M1 to M7) and gill slits (G2 to G9). G1 becomes the mouth and G2 is present only in the larva. In the shark the supporting skeleton of G1 is modified to form the biting jaws, G2 remains as the spiracle, whereas D6 to D8 become vestigial and two gill slits as well as M1 and M2 disappear. (MODIFIED FROM GOODRICH)

The nerves that emerge from the brain are called *cranial nerves*, while in the region of the trunk the nerves arising from the nerve cord (*spinal cord*) are called *spinal nerves*. In *Amphioxus*, the dorsal nerves have no ganglia and their cell bodies usually lie within the spinal cord, but in the vertebrates the sensory neurons that transmit impulses from receptors to the CNS by means of axons have their cell bodies in ganglia (*dorsal root ganglia*) just outside the CNS.

Although the general pattern of the lamprey's nervous system exhibits a great similarity to that of other vertebrate groups, some features of the nerves show the primitive conditions of the lamprey.

PRIMITIVE CONDITIONS OF THE LAMPREY'S NERVES
- The dorsal and ventral roots of the spinal nerves do not join together as they do in higher vertebrates.
- The nerves are simple and covered very little by insulating material (conditions known as *non-myelinated*).
- The *autonomic nervous system* is only slightly developed, controlling the muscle fibres of the gut and of the arteries.

In origin, the brain case of the vertebrates is made up of two parts. The first brain case was formed by the fusion of the cartilages that developed on the floor of the head region and the capsules that developed to protect sensory centres (nasal and auditory). To this the second part was added: elements of the gill skeleton were modified and became part of the skull.

Most cartilages were later replaced by bones and the brain case was further reinforced by additional bones.

The skeletal support of the gills forms an arch encircling the pharynx. Unlike the skeleton of the brain case proper, these *visceral arches* have a segmental arrangement. The foremost one, the *mandibular arch*, becomes the jaws to support the mouth in cartilaginous fishes. The second arch, the *hyoid arch*, was also modified when gills disappeared from this segment. The third visceral arch therefore formed the first *branchial arch* to support the foremost gills in fishes.

The segmentation of the head region illustrates origins of the cranial nerves and myotomes. The general pattern is schematically shown in Fig. 59 and tabulated below.

HEAD SEGMENTS

Segment	Somites	Cranial nerves	
		Ventral nerve root	Dorsal nerve root
1 (premandibular)	eye muscles: rectus superior rectus internal rectus inferior oblique inferior	oculomotor (III)	trigeminal: profundus ophthalmic (V1)
Visceral arches			
2 mandibular	eye muscle: oblique superior	trochlear (IV)	trigeminal (V 2 & V 3)
3 hyoid	eye muscle: rectus external	abducens (VI)	facial (VII) & acoustic (VIII)
4 first branchial of fishes	first myotome of *Petromyzon*	supplies first myotome	glossopharyngeal (IX)
5 second branchial	second myotome	supplies myotome & hypobranchial muscle	vagus (X 1)
6 third branchial	third myotome (first myotome of higher vertebrates)	,,	vagus (X 2)
7 fourth branchial	fourth myotome (second myotome)	,,	vagus (X 3)
8 fifth branchial	fifth myotome (third myotome)	,,	vagus (X 4)

GENERAL PATTERNS OF THE NERVOUS SYSTEM

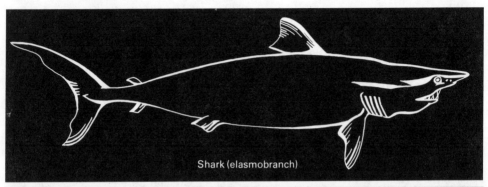

Shark (elasmobranch)

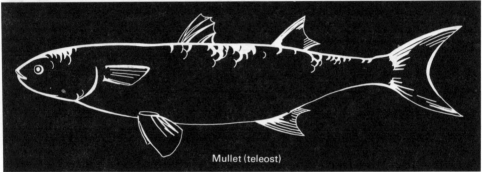

Mullet (teleost)

In the case of the vertebrates with jaws, the nervous system is well developed and, throughout the higher vertebrates, shows the same general pattern of three subdivisions: central nervous system (CNS), peripheral nervous system, and autonomic nervous system.

CENTRAL NERVOUS SYSTEM (BRAIN AND SPINAL CORD)

The brain (Fig. 60) may be described in five parts: the forebrain is divided into the end-brain (*telencephalon*) and the between-brain (*diencephalon*), together called the *prosencephalon*; the midbrain remains as one region (*mesencephalon*); and the hindbrain becomes *metencephalon* and *myelencephalon*, together called *rhombencephalon*.

The telencephalon is characterized by its connection with the *olfactory bulb* on each side. In fishes, the *cerebral hemispheres* (formed from the lateral walls of the telencephalon) are dominated by the olfactory connections, known as the *olfactory lobes*, which are fused in the median plane.

100

The beginning of the diencephalon is marked by a transverse fold in the roof, called the *velum transversum*. The walls of the diencephalon are thickened with masses of nerve cells forming the *thalamus* on each side. In higher forms, its main function is to co-ordinate the afferent sensory impulses from the brain-stem and spinal cord and relay this information to the cerebral hemispheres. These sensory impulses are mainly exteroceptive from the surface of the body and proprioceptive from muscles, ligaments, etc., so that the thalamus may be considered as a somatic sensory centre. It is also connected to a well-developed mass of nervous tissue in the floor of the diencephalon, the *hypothalamus*, to which the thalamus also relays impulses for visceral and homeostatic reactions. The floor of the diencephalon is depressed to form the *infundibulum* to which the pituitary body is attached. The roof is thin and bears the *epiphysis*, which is the vestige of the pineal eye—no longer a nervous structure. The cavity of the forebrain is called the *third ventricle* (the first and second are cavities of the cerebral hemispheres known as *lateral ventricles* in the lungfish and higher vertebrates), but the cavity of the mesencephalon is very much reduced and is known as the *aqueduct of Sylvius*.

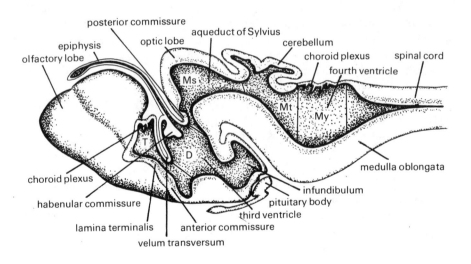

FIG. 60: Longitudinal section through the brain of the dogfish *Scyllium* (FROM DE BEER). T: telencephalon, D: diencephalon, Ms: mesencephalon, Mt: metencephalon, My: myelencephalon

The mesencephalon has a thick floor with fibres, *cerebral peduncles,* connecting the cerebral hemispheres to the spinal cord; the thick sides serve as the locomotive centre and pathways to the thalamus, and the thick roof, the *tectum,* consists of the paired optic lobes (*colliculi*) functioning as reflex centres of vision and hearing. The roof of the metencephalon is thickened to form the *cerebellum,* which receives impulses from all parts of the body, particularly from proprioceptive receptors, and controls muscle action. The myelencephalon (or *medulla oblongata*) has a thin roof and its cavity is known as the *fourth ventricle.*

The brain is surrounded by a membrane (*pia mater*) which carries blood vessels and dips down in folds from the roofs of the third and fourth ventricles to form the *choroid plexus.* Outside the pia mater there is another membrane, the *dura mater,* which has a protective function and is applied to the inner wall of the skull. There are only few commissures connecting one side of the brain with the other by tracts of nerve fibres. Of these, the *habenular commissure* and the *posterior commissure* are found in the roof of the diencephalon and mesencephalon respectively. There is also an *anterior commissure* (or *transverse commissure*) in the *lamina terminalis* (the anterior end of the brain proper).

The myelencephalon passes back gradually into the spinal cord, which has thick walls and a reduced central cavity. The nerve cells are grouped round the centre of the cord and form the *grey matter.* Outside the grey matter there are two tracts of nerve fibres forming the *white matter.* This arrangement also occurs in the brain, which has central grey matter as *cortex* and peripheral white matter as *medulla.* However, the grey matter is found outside white matter in the optic lobes and cerebellum.

The spinal cord is encased and protected by the neural arches of the vertebral column, through which it runs as far as the tail in most vertebrates. In man, the vertebral column has become longer than the spinal cord during evolution, and the spinal cord does not extend into the lower lumbar and sacral regions of the column. The spinal nerves supplying the lower sections of the trunk and legs originate in the upper lumbar region and travel down through the vertebral column forming a 'horse's tail'.

PERIPHERAL NERVOUS SYSTEM (CRANIAL NERVES AND SPINAL NERVES)

Cranial nerves (peripheral nerves emerging out of the skull) have roots similar to those of the spinal nerves; their origin is segmental (see p. 99). Four of them, however, do not show this pattern. These cranial nerves of special origin are the *nervus terminalis,* the *olfactory* nerve, the *optic* nerve and the *acoustic* nerve. The nervus terminalis is a very small nerve found at the anterior end of the brain. The olfactory nerve has no ganglion along its length. The optic nerve is formed from axons of the ganglion cells that are distributed to the retina of the eye. Because the retina is originally an outgrowth from the diencephalon (see p. 94), the nerve is also part of the CNS. The acoustic (or *auditory* or *vestibulocochlear*) is an extra dorsal nerve of the third (hyoid) segment of the head; it may be considered as a special branch of the seventh nerve or as an anterior lateral line nerve, a possible remnant of the extra segmental nerves that might have existed at an early evolutionary stage but have since disappeared from other segments.

The rest of the cranial nerves fall into two groups according to their origin: (a) somatic motor nerves (ventral roots), and (b) branchial arch nerves (dorsal roots). Somatic motor nerves supply striated muscles which are developed from somites. Branchial arch nerves are primarily sensory nerves, but often contain motor elements.

CRANIAL NERVES

	Origin	Distribution	Function
CRANIAL NERVES OF SPECIAL ORIGIN			
Nervus terminalis	ventral end of lamina terminalis	mucous membrane of nasal sac	?
olfactory (first cranial nerve—I)	receptors of nasal sac	forms olfactory bulbs	olfaction
optic (II)	diencephalon	rods and cones of retina	sight
acoustic (VIII)	medulla oblongata	inner ear	sense of balance, audition, selective inhibition of auditory stimulation
SOMATIC MOTOR NERVES			
oculomotor (III)	ventral side of cerebral peduncles (mesencephalon)	eye muscles (see p. 99)	movements of eye inwards, upwards & downwards
trochlear (IV) (or pathetic)	dorsal side of mesencephalon	eye muscle (see p. 99)	movements of eye inwards & downwards
abducens (VI)	floor of metencephalon	eye muscle (see p. 99)	movements of eye outwards & upwards
hypoglossal (XII)	ventral side of medulla oblongata (in reptiles, birds & mammals only)	tongue	movement of tongue muscle

BRANCHIAL ARCH NERVES

trigeminal (V)	metencephalon	branches from *semilunar ganglion:* superficial ophthalmic, profundus ophthalmic, maxillary, mandibular (face, head, cavities of teeth, upper neck)	*sensory:* touch, temperature, pain, pressure, proprioception *motor:* jaw movement
facial (VII)	metencephalon (near V)	branches from *geniculate ganglion:* superficial ophthalmic (joins the same of V), buccal (joins maxillary of V), palatine, prespiracular, hyomandibular (hyoidean & mandibular), dorsal cutaneous; all in head region	sensation of ear skin and anterior taste buds of tongue (in higher forms); low-frequency vibrations in water through the lateral line system of head region (in fish and larval amphibians)
glossopharyngeal (IX)	medulla oblongata	first pair of gills (in lower forms); pharynx and tongue (in higher forms)	*sensory:* pharynx and posterior taste buds of tongue *motor:* control of pharynx and secretion from salivary glands
vagus (X)	medulla oblongata	gills, viscera and lateral lines of body in lower forms; pharynx, larynx, lungs and viscera in higher forms	*sensory:* ear lobes, lungs, heart, lateral line, pharynx, stomach, abdominal viscera *motor:* parasympathetic control of viscera
spinal accessory (XI)	medulla oblongata (in reptiles, birds & mammals only)	shoulder girdle, thoracic and abdominal viscera (joins a branch of X)	*motor:* some muscles of shoulder girdle, control of pharynx, larynx, palate and viscera

The hypoglossal nerve is not a cranial nerve in fishes and amphibians. The corresponding nerves in fishes emerge from several ventral roots of the spinal cord (hence spinal nerves) and run down to the ventral side of the body to innervate the hypobranchial muscles (see Fig. 59, p. 98). In

Plate I   Playground and bowers of bower-birds

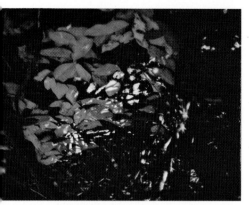

(a) Tooth-billed bower-bird *Scenopoeetus dentirostris*

(b) Spotted bower-bird *Chlamydera maculata*

(c) Great bower-bird *Chlamydera nuchalis*

(d) Yellow-breasted bower-bird *Chlamydera lauterbachi*

(e) Satin bower-bird *Ptilonorhynchus violaceus*

(f) Golden bower-bird *Prionodura newtoniana*

Plate 2(a)

Plate 2(b)

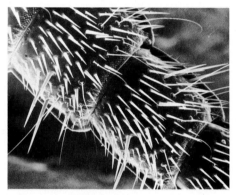

Plate 2(c)

Plate 2(d)

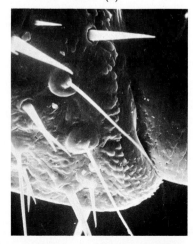

Plate II Cercus of the cockroach *Periplaneta americana* photographed with the scanning electron microscope. See also Fig. 89.

(a) Cockroach. (b) Whole cercus magnified 12 times. The smaller appendage at middle right is a style, found only in the male. (c) Portion of cercus magnified 60 times. Stout bristles and long sensory hairs can be seen.

(d) Magnification 300 times. The delicate sensory hairs have a complex socket at their base. These hairs respond to air movements and low frequency sounds.

amphibians it is the first spinal nerve to come out behind the skull, and runs to the muscles beneath the tongue which depress and elevate the pharyngeal floor for breathing.

The spinal nerves connect the receptor and effector organs of the body limbs with the spinal cord. They contain both sensory and motor elements as the dorsal and ventral roots join together on each side of each segment (see p. 34). In the grey matter of the spinal cord they form reflex arcs by synapsing with interneurons. They are also connected with other inter-neurons running in ascending or descending tracts inside the spinal cord.

AUTONOMIC NERVOUS SYSTEM

The autonomic nervous system (Fig. 61) consists of two divisions, the *sympathetic* and *parasympathetic*, and their actions are usually antagonistic to each other. All important organs receive innervations from both divisions. Stimulation of sympathetic nerves tends to increase the activity of an animal, speed up circulation and slow down digestion, while the action of the parasympathetic tends to slow down activity and promote digestion. Thus, heartbeat, blood pressure, the activity of smooth muscles of numerous visceral and pelvic organs, and the secretion of certain endocrine glands are controlled by the autonomic nervous system. No autonomic system is known in *Amphioxus*, though autonomic functions exist. In the lamprey both the sympathetic (though poorly developed) and parasympathetic are known to exist. In sharks (cartilaginous fish) the system does not extend into the head region, but in the bony fish and other higher vertebrates the co-ordination and control of the autonomic functions are found in special centres in the hypothalamus and medulla oblongata as well as in the spinal cord. The nerve fibres of the sympathetic system are not myelinated.

In the autonomic nervous system, one nerve connects the brain or the spinal cord with a ganglion (*preganglionic nerve*), and another nerve runs from the ganglion to the muscle or glands (*postganglionic nerve*). The muscles innervated by these nerves are always smooth (involuntary), and for innervation at least two nerve cells are involved. Striped (voluntary) muscles are innervated directly from the brain or the spinal cord by nerve cells which run all the way to muscles without ganglia. The synaptic connections, necessary for the co-ordination of stimuli or the inhibition and determination of responses, are contained in the autonomic ganglia as well as in the central nervous system.

The sympathetic system is composed of a chain of ganglia along the vertebral column, each segment basically having a separate sympathetic ganglion. Each ganglion, in the thoracic region and in the first few segments of the lumbar region, receives a nerve from its corresponding spinal nerve. This connecting nerve is called the *ramus communicans*. The sympathetic

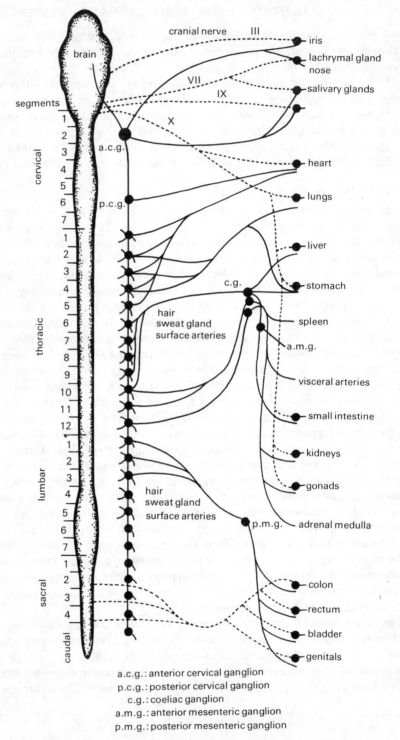

FIG. 61: Distribution of the autonomic nervous system (MODIFIED FROM WOODBURNE): sympathetic system (unbroken lines) and parasympathetic system (broken lines)

chain extends into the head region in front and the sacral region in the rear without rami communicantes.

The parasympathetic system originates in two regions of the central nervous system: one portion in the cranial and the other in the sacral region. The cranial portion is represented by the cranial nerves III, VII, IX and X, whereas the sacral portion arises from three segments of the sacral region and supplies the pelvic organs. Unlike the sympathetic system, the parasympathetic has no chain of ganglia but has ganglia lying close to the gland or muscles to which the nerve is distributed.

In order to maintain homeostasis (constancy of internal environment), the animal must be able to control and co-ordinate the activities of all organs and systems into a harmonious pattern. This is achieved through both nervous co-ordination (by the central nervous system and the autonomic nervous system) and chemical co-ordination (by the endocrine glands), the two interacting either directly or indirectly. For example, the pituitary gland, the so-called 'master gland' of the endocrine system, is strongly influenced by the nearby hypothalamus, and some of its hormones are even produced by ganglia in that region of the brain. Also, the adrenal medulla, an endocrine organ, is composed of modified nerve cells which secrete adrenalin under the control of the central nervous system. This control is achieved by the sympathetic system relaying nerve impulses from the brain. Thus emotions such as fear or anger can cause a high-level secretion of adrenalin into the bloodstream and prepare the body for extra muscular exertion.

HIGHER VERTEBRATES

The major breakthrough in the advancement of vertebrates above the level of Pisces involved the evolution of three important features by which they emancipated themselves from their dependence on the aquatic environment. These features developed in steps, and each step is represented by some living forms.

At *first*, the swim-bladder (a hydrostatic organ) of the bony fish developed into an accessory respiratory structure. Since the swim-bladder is a derivative of the alimentary canal, this respiratory lung is derived originally from tissues of the gut. With the respiratory lung, the lungfish (order Dipnoi) can breathe air if the environment is polluted or dries temporarily. The lungs of the terrestrial vertebrates have a similar origin in development.

Most amphibians metamorphose from the aquatic larva to the adult adapted to the terrestrial or semi-aquatic environment. Here the *second* step towards terrestrial life is seen in the locomotive organ, which consists of two pairs of limbs in place of fins. Hence, amphibians and all higher

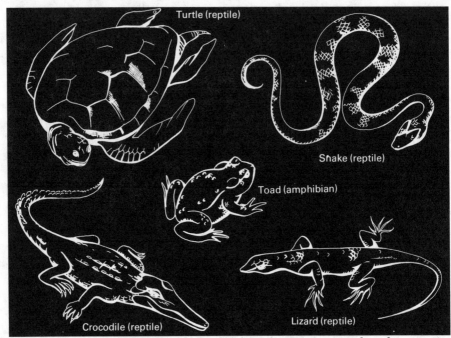

Turtle (reptile)

Snake (reptile)

Toad (amphibian)

Crocodile (reptile)

Lizard (reptile)

vertebrates are typically land animals (though some returned to the aquatic environment later, showing secondary adaptation), and are collectively called the tetrapods (*Tetrapoda*). The limbs basically possess five digits (i.e. are *pentadactyl* in condition), and locomotion on land is much facilitated.

Since the eggs develop in the water and the larvae respire by means of gills, the invasion of dry land was still limited. The *third* step, then, was to produce eggs which could develop on land. Some early amphibians developed an egg of this type, covered with shell, in which an aquatic environment was maintained for the embryo to develop. In such a group, the eggs had to be fertilized internally by copulation. The egg shell was horny or calcareous to serve protective functions, but at the same time permeable to allow gas exchange. It contained fluid in the *amnion* (extra-embryonic ectoderm). In other words, the *amniotic cavity* developed to protect the embryo from desiccation. Another foetal membrane, the *allantois*, was also developed from the bladder of amphibians (originally an outgrowth from the rectum) to carry blood vessels and assist in respiration. Metabolic wastes were excreted by the embryo and stored in the *allantoic cavity*. The nutrients necessary for the development of the embryo were contained in the yolk sac. Thus these eggs became relatively independent of the environment. All tetrapods that produce this type of egg are called the amniotes (*Amniota*), consisting of Reptilia, Aves and Mammalia.

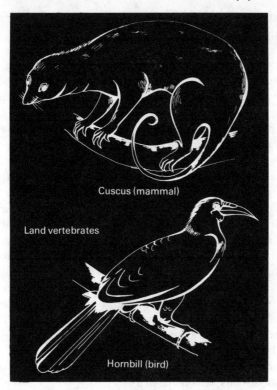

Cuscus (mammal)

Land vertebrates

Hornbill (bird)

There are associated changes in other organizations of the body at each step of advancement. Primitive conditions of the lower forms are apparent when comparison is made with the higher forms, but at the same time there are other features that are unique to a particular group, indicating specialization at that level. In considering the evolutionary origin we speak of certain organs as *primitive*, *specialized* or *degenerate*, according to the degree of development of the organ in question. In the comparative approach, the primitive conditions do not necessarily mean unspecialized (or generalized) conditions, but are ancestral features indicated by (a) the lack of development of structures, (b) the prototype features of structures, and/or (c) the primitive mode which is replaced by more efficient modes in higher forms. The first two show some directional change through evolution, while the last is unique in its function at that level of organization. The structures associated with this unique function may become vestiges, disappear altogether, or be modified to serve other functions in higher forms. When the same function is performed by different organs in higher forms, such organs are said to be *analogous* (e.g. gills of fish and lungs of tetrapods), whereas if different functions are performed by organs derived from the

109

same original structure, they are said to be *homologous* (e.g. the swim-bladder of fish and lungs of tetrapods).

Directional changes from the prototype to complex structures are illustrated by the cerebral hemispheres, which are formed by a pair of anterior extensions of the third ventricle in the diencephalon. The lateral ventricles (the first and second ventricles) thus formed in the telencephalon communicate with the third ventricle through the *foramina of Monro*. The cerebral hemispheres developed at the level of the lungfish and gradually increased in size through amphibians, reptiles, birds and mammals. At the same time their cavities (the lateral ventricles) were reduced. The increased size of the cerebral hemispheres in reptiles and birds is due to the enlarge-ment of the *corpus striatum*, which is considered to be responsible for co-ordinating complex innate behaviour, whereas the enormous cerebral hemispheres in mammals are due to the *cerebral cortex* (superficial grey matter). They extend posteriorly and cover the diencephalon and mesen-cephalon. A number of lobes marked by fissures and folds show increasing complexity from lower to higher mammals. The *neocortex* (the new cortex) that is developed in amniotes has well-defined areas relating to the sensation and movements of localized regions of the body. By electrically stimulating various parts of the neocortex, it is possible to determine the local represen-tation of sensory and motor areas in the cortex. Such representation is more defined in higher forms. This evolutionary process is called the *corticaliza-tion*; it is accompanied by the enlargement of the association areas of the cortex and is most conspicuous in primates.

Other examples of structures that show increasing complexity in the course of evolution include the heart, the vertebral column, the endocrine system and the associated nervous system. Increased complexity of these organs in higher vertebrates usually indicates increased efficiency and effectiveness in their respective functions.

The changes of mode are seen in the accommodation of vision, the method of breathing, the body cover for protection and insulation, and many other features which show parallel development with major steps of evolution. For example, the skin of amphibians is naked and semipermeable serving as a respiratory surface. It is imperfectly protected from predators and heat by mucus secreted by glands. In reptiles, birds and mammals the body surface is covered with horny scales, feathers and fur respectively.

Specialization of structures is often associated with the zone of adaptation into which the animal entered, and this is also seen in the details of sensory capacities and nervous organization. For example, fish (particularly nocturnal species) have a well-developed olfactory nerve, which is reflected in the large olfactory lobes in the telencephalon. Frogs have an acute sense of hearing, and this is associated with accessory structures related to the reception, amplification and conduction of vibrations from the external environment. The middle ear has the *tympanic membrane* (eardrum), and between this and the outer wall of the auditory capsule is a considerable space. This communicates with the pharynx by the short *eustachian tube*. The connection of the tympanic cavity with the pharynx helps to maintain equality of pressure on each side of the tympanic membrane. During the breeding season hearing is very important, and they can detect 50 to 10,000 vibrations per second. On the other hand, the olfactory organ of frogs is simple and quite unimportant. In birds the visual sensory information becomes dominant and the centres of vision are well developed in the optic lobes. Birds have very keen sight and accommodation for distance is accomplished by both lens and cornea with extreme rapidity; this meets the exacting needs of flight. The eyes of predatory birds are located on the same plane as man, but those of hunted animals are located at the sides of the head. In the latter species, each eye can see different objects as they possess *monocular vision* to cover a much wider angle than if the eyes were trained on the same object. However, the fields of vision of the two eyes overlap in front. In addition, birds have excellent colour vision and light discrimination as well as dynamic vision. The ear, though not externally apparent, is also well developed in birds, and in general the range of hearing is in the high frequency of sound. On the other hand, the sense of smell is poorly developed except for the kiwi and some sea birds. Associated with the development of the corpus striatum, birds have a very highly developed system of innate perceptory patterns of behaviour. The equilibrium required for the dynamics of flight is centralized in the co-ordinating centres of the cerebellum, which is greatly increased both in size and in the extent of fissures.

In mammals many unique features are found in the brain (Fig. 62). The functional tie between the two hemispheres of the brain is much increased. The two sides of the cerebellum are connected by a band of nerve fibres known as the *pons* (*pons Varolii*), which carries impulses to the cerebellum for co-ordination of voluntary and involuntary movements of muscles. The cerebellum is solid and the fourth ventricle does not enter it. The optic lobes have a transverse furrow dividing each lobe into two bodies, forming the *corpora quadrigemina*. The sides of the diencephalon are thickened, and the optic thalami of two sides are connected across the third ventricle by the *soft commissure* (*interthalamus*), as well as by the habenular commissure near the pineal body. The anterior commissure connects two halves of the corpus striatum on the floor of the telencephalon, while the posterior commissure is in the roof of the mesencephalon. The cerebral hemispheres are mostly composed of grey matter (*cerebral cortex*) and the *corpus callosum* connects the two sides across the median fissure by a broad band of

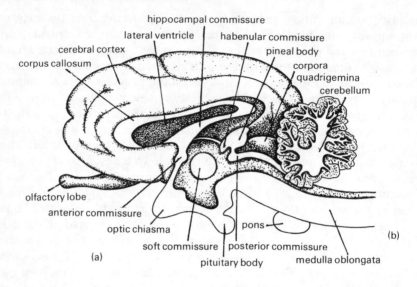

(a)

(b)

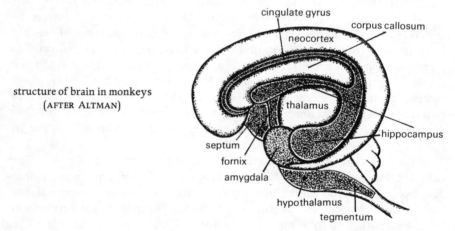

structure of brain in monkeys
(AFTER ALTMAN)

transverse fibres. This allows the two hemispheres of the cortex to act as a single unit (e.g. in co-ordination of the two visual fields.) On the floor of the lateral ventricle there is a semicircular ridge, the *hippocampus*, and along its anterior ridge another ridge of fibres forms the *taenia hippocampi*. Two taeniae unite in the olfactory region of the hemispheres to form a median longitudinal strand called the body of the *fornix*, which becomes the *hippocampal commissure* below the corpus callosum. These connections in the cerebral hemispheres form an important part of the *limbic system*, which is largely subcortical. The limbic system is concerned with motivational and emotional activities, in contrast to the strictly informational aspects of cortical function.

A rather diffuse network of nerve cells and fibres, known as the *reticular formation*, extends from the medulla to the mesencephalon through the pons and enters the thalamus to form the *reticular nuclei*. It is known to project into many areas of the cortex for the arousal reaction (Fig. 63a). The hypothalamus has mediating functions for emotions and motivations through its connection with the autonomic nervous system. The physiological factors controlling the excitatory and inhibitory functions of the hypothalamus are shown in Fig. 63b.

The pia mater with its blood vessels projects into the lateral, the third and the fourth ventricles, forming a choroid plexus in each. The choroid

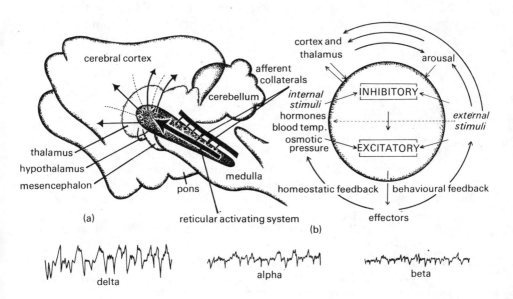

(a)

(b)

delta   alpha   beta

FIG. 63: Some functions of the brain. (a) Reticular activating system into which the major sensory pathways send collaterals and cause arousal (AFTER STARZL, TAYLOR AND MAGOUN). The electroencephalogram (EEG) is used to measure such activity of the brain. The electrical activity of the brain is less than 50 microvolts but can be amplified several million times when picked up by the electrodes placed symmetrically on both sides of the head. Different brain wave patterns in various parts or states of the cortex can thus be identified in the EEG. Alpha waves have rhythmic activity of 10 cycles per second, beta waves of 20 to 30 per second or more, and delta waves of less than 5 cycles per second with relatively large amplitude. Delta waves are characteristic of sleep, but if the reticular formation is stimulated these waves change to alpha or beta waves, characteristic of arousal or excitement.

(b) Factors controlling the excitatory and inhibitory functions of the hypothalamus.

plexuses secrete *cerebrospinal fluid* which fills the space (*subarachnoid space*) between the pia mater and the *arachnoid membrane* as well as the canals of the spinal cord and the ventricles of the brain. This fluid acts as a cushion around the brain and spinal cord and at the same time helps to remove metabolic wastes from the nervous tissue by the continuous secretion and resorption of the wastes into the bloodstream. The protective dura mater and these membranes are called the *meninges*.

Another important feature of mammals is the fact that the viscera are completely divided into two parts by a transverse partition called the *diaphragm*. In front of the diaphragm the thoracic region contains lungs in the *pleural cavities* and the heart in the *pericardium*. Behind the diaphragm the abdominal region contains the digestive system. The diaphragm is pierced by the alimentary canal and major blood vessels. The muscles of the diaphragm are of somatic origin and are innervated by the *phrenic nerves*, formed by the fourth and fifth spinal nerves and running on each side of the heart to reach the diaphragm. The movements of the diaphragm are largely responsible for the intake of air during breathing.

SUMMARY OF THE FUNCTIONS OF THE CNS

The spinal cord:

- The *reflex arc* (receptors—sensory neurons—interneurons—motor neurons—effectors) integrates reflex behaviour.
- The ascending spinal columns conduct impulses to the thalamus and also contribute collateral nerves to the reticular formation.
- The descending spinal columns conduct impulses from the cerebral cortex.
- The spinal nerves are connected to the sympathetic system by means of rami communicantes.
- The sacral portion of the parasympathetic system originates in the spinal cord and supplies pelvic organs.

The brain:

- The ascending and descending connections with the spinal cord have centres of sensory, motor and association structures.
- The medulla has the reflex arcs for many cranial nerves and controls respiratory and cardiovascular functions in lower forms.
- The cerebellum has sensory centres and co-ordinates movements for balance.
- The thalamus receives the reticular formation from the mesencephalon and co-ordinates afferent sensory impulses. It relays information to the cerebral hemispheres and to the hypothalamus.
- The hypothalamus controls the autonomic nervous functions and stimulates the pituitary gland.
- The neocortex in amniotes has well-defined areas representing sensation and the movements of localized regions of the body.

- The *palaeocortex* (the old cortex) in higher forms becomes subcortical and forms part of the limbic system.
- The limbic system, originally concerned with olfaction, connects the hippocampal, the hypothalamic and the thalamic regions with the cortex in a circuit (*Papez circuit*), and under chemical stimulation it directs innate behaviour patterns. However, different chemicals induce different behaviour in different species. The internal drives may be generated in this circuit according to various inputs from the reticular system and the hypothalamus, as well as from the neocortex in a feedback system. Messages may be circulated through this system after the input ceases to enter the circuit.

CAUSATION OF BEHAVIOUR

The word 'causation' means 'causal relationships' underlying mechanisms of behaviour and does not mean 'causes' of behaviour. Hunger, for example, may be a necessary cause for nutrition behaviour, but not a sufficient cause. In order to explain the act of eating causally, the external stimuli relating to this act must be analyzed in terms of the internal state of hunger. Then, hunger may not even be a necessary cause, as there are cases in which animals show the act of eating without being hungry. They may be socially facilitated by seeing other animals eat or they may eat in trying to thwart other drives. Studies of these causal relationships reveal that even such simple behaviour as eating cannot be explained in terms of simple causes, because the causal factors interact to inhibit or facilitate elicitation of behaviour through nervous and chemical co-ordination mechanisms.

As we have seen, anatomical contiguity does not necessarily indicate that the functional systems are also contiguous in different animal groups. Therefore, the models of physiological mechanisms derived from the study of one group of animals may not apply to other animal groups. Moreover, the functional models based on the study of behaviour may represent different mechanisms in different animals? When a particular behaviour pattern is obtained by stimulating various parts of the brain electrically or chemically, models of physiological mechanisms of behaviour may be constructed. The fact that the structure of such models is often different from that of functional models based on the study of behaviour indicates complexity of behaviour mechanisms. In recent years some attempts have been made to relate functional models to neuro-anatomical features of the animal in order to narrow such gaps of knowledge.

One of the successful attempts is found in the model suggested by J. R. Smythies, who tried to simulate the major routes of information flow in the brain with circuit operations of the computer process (Fig. 64). In his model,

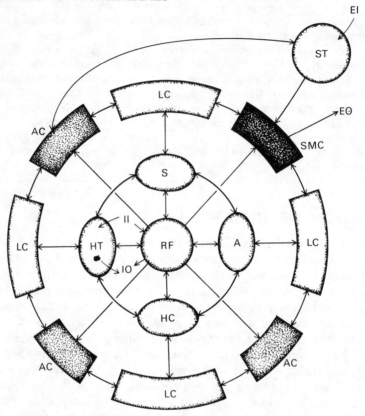

FIG. 64: A circuit model of subcortical functions (AFTER SMYTHIES).
A: amygdala, AC: association cortex, EI: external input, EO: external output, HC: hippocampus, HT: hypothalamus, II: internal
input, IO: internal output, LC: limbic cortex, RF: reticular formation, SMC: sensori-motor cortex, ST: specific thalamic nuclei

information about the external environment is fed continuously in a coded
and analyzed form to the hippocampus, and then circulated round the
limbic circuits in the form of an internal representation of the external
world. The *amygdala* feeds into these same circuits information about the
internal environment. This 'record of experience' is fed to the midbrain
reticular formation and thence may be fed, suitably modulated, to the
thalamic reticular formation as a programme for motor behaviour, and
to the hypothalamus as a correlated programme for visceral concomitants of
behaviour (emotion). The 'tape' produced by the analyses of incoming
information can act as the programme for the executive mechanism of
the reticular formation which integrates these manifold and complex
programmes.

116

6   Stereotyped Behaviour

The control systems that we have studied provide the basis of animal behaviour. In the rest of the book we shall examine mechanisms and functions of behaviour at the individual and social levels. The old controversy of 'nature' versus 'nurture' has been restated in modern times: how do inheritance and environment interact to shape behavioural characters? We begin our study by examining distinct types of innate and learned behaviour.

REFLEXES

In contrast to the behaviour of humans where many responses are very much dependent on past experiences, behaviour in less highly organized animals involves a large proportion of stereotyped acts.

A fly landing on a table top usually moves around, turning occasionally. It may stop while the antennae, head or legs are cleaned, and this may be followed by more locomotion or possibly flight. After a period of study an observer may reach the stage where he knows what to expect next in the fly's behaviour, and also how the particular acts will be executed. If we stop whatever we are doing, *our* next act could be very hard for an observer to predict. Instead of moving towards the dining room because we are hungry, we may undertake some other small activities before the stimulus of an empty stomach directs us towards food. If we have decided to restrict our diet, even though hungry and stimulated by the presence of food, we still may not eat. A starved fly always accepts offered food.

These differences in behaviour can be related to differences in ability. The fly has very little ability to foresee the probable future course of events, whereas we have this ability and it colours our behaviour accordingly. In general, the further down the scale of animal organization we go, the more stereotyped their activities become and the less 'free will' is evident.

The fly maggots in a rotting animal carcass are just about to pupate. If we direct a bright light onto a maggot's head, most and probably all of the individuals tested will turn and, swinging their heads from side to side, move away from the stimulus. A cockroach will run away at high speed if a puff of air is directed at its hind end (cerci). Newly-hatched freshwater

117

shrimp larvae all congregate at that side of the aquarium nearest the window because they move towards the light.

All of these responses are simple, innate, stereotyped and predictable. The stimulus is received, the animal responds, and to the observer the behaviour appears rigid and forced; the animal is behaving as if it were a machine. But it is never quite as simple as this, because all behaviour is very dependent on the internal state of the animal responding. Age, circulating hormones, sensory information from internal receptors and many other factors influence the type of response an animal gives to a particular stimulus. In addition, an animal often responds to the total stimulus field rather than one stimulus in isolation. To a hungry octopus, the sight of a crab is a powerful stimulus. The octopus will glide towards its prey, then pounce and bite it, subduing it with toxin as it does so. But the crab must be moving along the bottom and not dangling from a string held by an experimenter. In this latter case the stimulus is very similar—but the octopus responds by exploring the object it has seen; a few jets of water are often squirted at it, and only after this cautious approach is the crab attacked.

Early in this century it was common for physiologists to attempt an explanation of all behaviour in terms of reflexes, and animals whose behaviour showed simple stimulus-response systems gave weight to their arguments. Complex behaviour patterns were explained by postulating chain reflexes: one sequence of the pattern triggered the next, it having been triggered by the preceding one. While it is easy to give examples of simple reflex activities figuring prominently in behavioural acts, it is now appreciated that simple reflexes are physiological abstractions. The classical reflex arc of receptor, sensory neuron, interneuron, motor neuron and effector never exists in isolation. Between receptor and effector many different factors may operate. Each reflex arc is potentially connected, via interneurons, to all other neurons and arcs; hormones might operate to direct impulses along alternative motor pathways; the reflex act might be triggered by a removal of inhibition from certain pathways, and so on. The reflex cannot even be considered as the basic functional unit of the nervous system, because many behaviour patterns can arise spontaneously in the absence of triggering stimuli, and still other patterns rely on activity from the CNS to maintain and complete them after an initial trigger from a receptor.

The reflex, as currently understood, of stimulus → intervening variables → response, may be quite simple or extremely complex. Any attempt to classify reflexes in steps of increasing complexity must inevitably be artificial. Because the same type of act performed by different animals involves different components, considerable overlap occurs between the

118

categories and only a few reflex acts are understood in any detail. With these limitations accepted, a classification of reflexes is presented here; it is considered useful *only* as an aid to description.

PRE-NERVOUS REFLEXES

All definitions of reflex arcs and most definitions of reflex activities include reference to the functioning of the nervous system and its components. But in the broad sense of a reflex act being an animal's stereotyped response to a stimulus, protozoans and sponges which lack nervous structures also show this type of behaviour.

The stimulus received by *Paramecium* bumping into an object while swimming leads to a response: reversal of ciliary beat, causing a backward movement (see Fig. 10, p. 27). The osculum of sponges will slowly close in response to such stimuli as exposure to air during the falling tide, exposure to extremes of temperature, or injurious chemicals.

TWO-COMPONENT REFLEXES

Within the coelenterate group of animals, reflex pathways can be found which involve receptor and effector only; the interneuron of the classical reflex arc may be absent.

In the body wall of *Hydra* the contractile elements are found in the bases of special epithelial cells (*musculo-epithelial cells*). The sensory cells scattered in between the epithelial cells have processes which may connect synaptically either with nerve cell processes or with the contractile cells. The reflex pathway in the former case is 'typical', namely receptor → interneuron(s) → effector, but in the latter the receptor links directly with an effector, i.e. only two components are present.

MONOSYNAPTIC REFLEXES

When testing the knee jerk reflex, the doctor chooses a fairly simple reflex pathway. The hammer taps the tendon beneath the kneecap, which stretches the attached muscle, and stretch receptors within the muscle are excited and send information via sensory nerves (called 1A *afferents*) to the spinal cord (Fig. 65a). The sensory nerve here makes direct synaptic connection with motor neurons (*alpha motor neurons*) which, when they fire, cause the muscle they innervate to contract (and the leg is flexed).

The *major* pathway involved thus has one synapse only, between receptor and effector neurons. But the sensory neuron will also be making many other synaptic connections with interneurons. The monosynaptic reflex arc therefore does not exist as a separate entity, and this is true of

119

all reflex pathways in highly organized animals. The sensory nerve bringing information from a receptor breaks up into a number of endings, and while some endings may contact motor neurons (thus giving a mono-synaptic pathway), many others contact interneurons which contact other interneurons and so on (Fig. 65b).

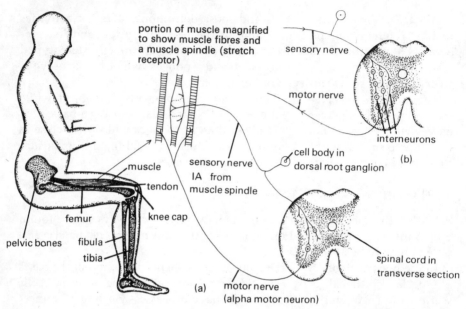

portion of muscle magnified to show muscle fibres and a muscle spindle (stretch receptor)

sensory nerve

motor nerve

interneurons

muscle

tendon

sensory nerve IA from muscle spindle

cell body in dorsal root ganglion

(b)

femur

knee cap

pelvic bones   fibula

tibia

spinal cord in transverse section

(a)   motor nerve (alpha motor neuron)

FIG. 65: Reflex pathways: (a) neural pathway involved in the knee jerk reflex, and (b) a polysynaptic pathway

POLYSYNAPTIC REFLEXES

If we flex or withdraw our leg in response to a painful stimulus, the reaction would probably be performed with much more vigour than the leg movement which followed the doctor's hammer tap. An increase in the amount of response with increasing strength of stimulation is called *irradiation*. We are all familiar with the type of reaction we give after touching a fairly hot object compared with that after touching a red-hot object.

When a limb is withdrawn from a painful stimulus, the original position is not regained immediately after the stimulus stops. This continuation of a response beyond the duration of the stimulus is termed *after-discharge*. One mechanism which allows after-discharge provides many interneurons in a reflex pathway, giving a polysynaptic reflex arc. Between the sensory neuron termination and the motor neuron, one or many interneurons may

Plate 3(a)

Plate III (a) A chiton *Acanthozostera gemmata* photographed at night on the beach rock of a coral cay. The animal will return to its homesite (the light-coloured, oval area in the photograph) by following the trail browsed through the algae on its outgoing trip. Chitons experimentally displaced from their trails often find their way home, possibly using a topographical memory.
(b) Bent-wing bats *Miniopterus schreibersii* in a cave. They raise the temperature of a cave chamber before giving birth to young in it.
(c) A formation of straw-necked ibises *Threskiornis spinicollis* in migration

Plate 3(b)

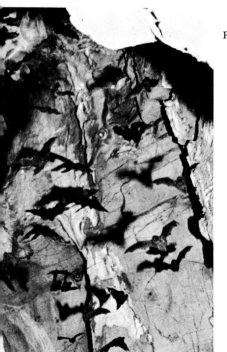

Plate 3(c)

Plate IV (a) Star finch *Neochmia ruficauda* (photo: C. H. Greenewalt)

Plate IV (b) Gouldian finch *Chloebia gouldiae* (photo: C. H. Greenewalt)

Australian finches are colourful and often kept in captivity. In the field closely related species often flock together but different markings on the head and body help species identification (see Fig..98).

Plate IV (c) A threat posture of the Queensland rock-wallaby *Petrogale inornata*. It lives in a colony associated with a rocky environment.

occur (Fig. 65b). There is a delay in impulse transmission at every synapse; thus activity in an effector can be maintained for a period while all the neural pathways are followed.

After-discharge, like irradiation, has important survival value. If a cockroach were to run rapidly only while receiving warning stimuli it would not evade predators very efficiently. The roach therefore continues to run for a period after the stimulus has ceased and this helps to move it outside the influence of the stimulus, often to the safety of a crevice. After-discharge of impulses, which are initiated by ganglia in the thorax and the brain and travel along the motor nerves to the leg muscles, keeps the roach running.

The above description of the polysynaptic reflex pathway following stimulation of a pain receptor is very much simplified, and although this pathway may be utilized, two reflex pathways are normally involved when we move our limbs in response to stimulation. The route of the impulses is as follows (Fig. 66b): the pain receptor fires and impulses pass along the sensory nerve to the spinal cord. Within the cord, synaptic connection is made with interneurons which contact the motor neurons (*gamma motor neurons*) that run to the ends of the stretch receptor in the muscle, called a *muscle spindle* (Fig. 66a). The ends of this spindle are contractile and under the influence of gamma neuron stimulation they shorten. This causes the stretch receptor mechanism in the middle of the muscle spindle to fire, and

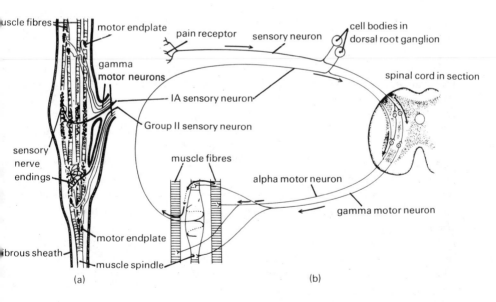

(a)   (b)

FIG. 66: (a) A vertebrate muscle spindle (FROM FLOREY), and (b) neural pathways following a painful stimulus and leading to a withdrawal response

impulses pass to the spinal cord via 1A sensory nerves. Again synaptic connection is made with interneurons, and this time contact is made with alpha motor neurons. These supply the muscle fibres and cause the muscle to contract.

When the amount of muscle shortening equals the amount of shortening initially suffered by the muscle spindle, the spindle is no longer under stretch and so ceases to initiate impulses. The muscle fibres, no longer receiving stimulation, relax. This double pathway of neural elements helps regulate the amount of muscle contraction. The information that is given to the muscle not only causes it to contract, but also governs how far it contracts.

It must be emphasized that this description of muscle contraction following stimulation is still very simplified. Within the muscle spindle other sensory endings may occur besides the central primary one (there may be from zero to five secondary sensory endings). Additional gamma motor neurons run to the spindle ends (from seven to twenty-five motor axons may innervate a single spindle). The roles of these additional components have not been considered.

MONOSEGMENTAL REFLEXES

Some reflex activities are confined to only one body segment. Such reflexes are easily seen in annelids or arthropods where the segmentation is distinct. The pathway involved may be mono- or multisynaptic.

In the earthworm *Lumbricus*, locomotion can occur even if the two longitudinal nerve cords that run the length of the body are cut. The co-ordinated movements are the result of intrasegmental reflexes. When the longitudinal muscles of the preceding segment contract, the stretch on the muscles of the segment in question is detected by receptor cells. Impulses are passed to the segmental ganglion along a sensory nerve, and within the ganglion synaptic contact is made with motor nerves running back out to the muscle. When these motor nerves are excited, the muscle contracts. Thus contraction of muscles in one segment triggers activity in the following segment and so on, producing a co-ordinated wave of contraction along the length of the worm.

In crabs or crayfish, reflex activity occurs in the limb of a single segment. If the inner surfaces of the large pincer-like claw are touched, the claw closes. The nervous pathway includes a thoracic ganglion. This reflex is easily demonstrated when a crab is out of water, but when under water the crab must be excited by chemical stimulation as well for the reflex to operate.

MULTISEGMENTAL REFLEXES

A reflex act may primarily involve one segment, but an influence may spread to other segments. A crab's eyes are positioned at the tips of long stalks. If strong stimulation is applied near the eye, the eyestalk is withdrawn. The appendages in front of the eye (the antenna and antennule), as well as the eye on the other side of the body, may also be moved.

In other multisegmental reflexes the segments may operate as a functional unit. Warning stimuli received at the head end of a crayfish give rise to impulses in the giant nerve fibres of the nerve cords. The giant fibres innervate the abdominal muscles whose concerted action produces an abdominal 'flick' forwards, darting the animal backwards.

POSITION REFLEXES

The brine shrimp *Artemia* swims upside down (at least, to *us* it swims upside down). Light stimuli are used to maintain this position, and we can test brine shrimps' reactions by placing a lamp beneath or at the side of a glass-walled aquarium. They will be seen swimming the 'right way up' or on their sides respectively (Fig. 67). The position reflex which allows the shrimps to maintain their basic position in space is an example of a *ventral light reaction*.

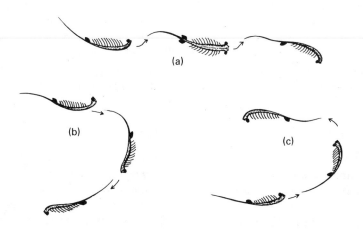

FIG. 67: The brine shrimp *Artemia salina* swims with its ventral surface towards the light regardless of the direction of the light. When the direction is changed from above to below, the shrimp responds to it by (a) rolling, (b) back somersaulting, or (c) front somersaulting. (FROM FRAENKEL AND GUNN)

123

THE BEHAVIOUR OF ANIMALS

Brine shrimps are unusual in positioning their ventral surface uppermost and also in their utilization of only one type of stimulus (light) to control this reaction. Other animals usually employ a number of different stimuli to orient themselves; tactile and gravitational stimuli are also used extensively. This ensures that in the temporary absence of some stimuli, such as light, the animal can still position itself correctly.

Swimming animals usually rely on gravity receptors to give information about their position, but if these receptors are removed, orientation to light is still possible. Crayfish, 'ordinary' shrimps and lobsters possess statocysts in their first pair of antennae. These are cup-shaped structures and contain heavy particles such as sand grains. The particles will change position within the statocyst as the animal is tilted. Sensory cells lining the statocyst wall detect their new position and give information for a correcting action to be made. When a shrimp moults, the lining of the statocysts and also the enclosed particles are lost. If we allow a shrimp to moult in an aquarium with only iron filings on the bottom, it will introduce the filings into its new, empty statocysts in the same way as it would use sand grains. Such a shrimp under the influence of a magnetic field will alter its position in space in relation to the altered positions of the filings. Statocyst removal does not prevent orientation; a dorsal light reaction enables correct body positioning.

PROTECTIVE REFLEXES

When warning stimuli of different types are received by animals, especially invertebrates, one reflex act is commonly shown and all other reflexes are temporarily inhibited. If an earthworm has partially extended out of its burrow, a touch, decrease in light intensity or ground vibrations can cause a rapid withdrawal. A scorpion receiving warning stimuli will elevate its tail and direct the sting forwards; a spider might freeze motionless.

CHAIN REFLEXES

A series of reflex acts, each triggering the next reflex, together produce a co-ordinated behavioural act called a *chain reflex*. The looping locomotion of leeches involves such a series of acts. When contraction of the *circular* muscles in the body wall extends the animal, the *longitudinal* muscles are reflexly inhibited. The anterior sucker attaches (this attachment cannot be made when the longitudinal muscles are contracted), and once this occurs the posterior sucker can detach. This triggers contraction of the longitudinal muscles, bringing the hind end forwards. The posterior sucker now attaches, and the cycle can be repeated.

Feeding in crabs is an example of a chain reflex under control of the brain.

Even if the brain is removed, an edible object presented to the claw is grasped, transferred first to the modified anterior legs and then to the mouthparts, and finally eaten if the taste is acceptable. The brainless animal will continue to accept, transfer and eat offered food even if its stomach is full; this can eventually lead to the stomach bursting!

REFLEXES AND THEIR IMPORTANCE IN CONTROLLING BEHAVIOUR

In the examples of behavioural acts described in this chapter, the animal's response depends on the prior receipt of stimuli detected by receptors on the outside of the body. These stimuli directly trigger the reflex response. Thereafter, in complex activities particularly where rhythmic patterns are involved, feedback from the periphery often controls muscular activity. A feedback loop exists which starts the next series of events before some central pacemaker within the CNS does. As we have seen, looping locomotion in leeches depends on peripheral control; the state of contraction of each of the muscle sets and the position of each sucker provide the necessary information to control the locomotory pattern.

Peripheral control of behaviour is not the only method available to animals, as central patterning is also commonly employed. The CNS gives rise to impulses, apparently in the complete absence of peripheral stimulation, and these serve to direct the behaviour to completion. Motor neurons to the swimmerets on the abdomen of crayfish continue to carry rhythmically patterned impulses, even when the abdominal nerve cord is isolated from the periphery. In contrast to looping locomotion, swimming in leeches depends on a pattern of impulses arising in the CNS.

Both types of control, central and peripheral, are involved in earthworm locomotion. In an earlier section (p. 122) it was shown that co-ordinated movements are still possible after the ventral nerve cords have been severed. Peripheral stretch receptors trigger intrasegmental reflexes. If a complete section of the body is cut away leaving only the ventral nerve cord intact (Fig. 68), a co-ordinated wave of contraction can pass along anterior and posterior sections as if no interruption was present. Locomotory movements can still occur in the absence of any peripheral stimuli which may have arisen in the removed segments. The existence of control systems, such as central patterning, makes the description of behaviour in terms of reflexes alone unsatisfactory. The reflex is only *one* of the many components which control animal behaviour.

Some behaviour patterns appear to arise spontaneously, in the absence of stimulation from the external environment. In the developing mammalian

125

embryo the limbs may show co-ordinated movements long before their reflex arcs have formed.

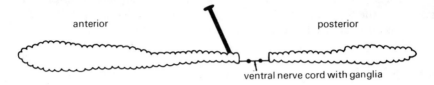

anterior   posterior

ventral nerve cord with ganglia

FIG. 68: In the earthworm co-ordinated locomotory waves (peristaltic waves) initiated anteriorly can pass posteriorly and this action is controlled either by peripheral stretch receptors triggering intrasegmental reflexes (see p. 122) or by impulses arising in the CNS. This latter control is demonstrated in the experimental work figured here. A portion of the earthworm's body was cut away leaving the nerve cord as the only connecting link between the two parts. Locomotory waves could still pass from the anterior to the posterior part. (AFTER WOOD)

The lugworm *Arenicola* (Fig. 69), which lives in U-shaped burrows on intertidal sand flats, shows a burst of feeding activity which subsides after every six or seven minutes. The feeding movements occur regularly even if the worm is denied access to any food by keeping it in a glass U-tube in an aquarium. These rhythmical outbursts of feeding activity are triggered by the oesophagus, which becomes spontaneously active every six or seven minutes even when removed from the body and kept in physiological salt solution.

Feeding behaviour in the anemone *Metridium* (see Fig. 42, p. 68) is followed by elongation of the column which persists for several hours. This behaviour can be triggered by introducing food juices into an aquarium containing an anemone, but the complex activities that follow can not be explained in terms of chain reflexes. For even when the water is drained and replaced with fresh seawater, thus removing any food stimuli, the feeding behaviour goes to completion. Moreover, when *Metridium* has been starved for some time, feeding activity can be triggered by inappropriate stimuli or can appear spontaneously without any apparent external stimulus.

DIRECTED REFLEXES

Many reflexes, such as leg withdrawal or eyestalk withdrawal, can be performed without regard to the animal's position in space. Normally, however, animals adjust their spatial relationships with respect to their source of stimulation; they perform *orientation reactions* or *directed reflexes*. An orientation reaction is a *steering* reaction and locomotion may or may not accompany such a response.

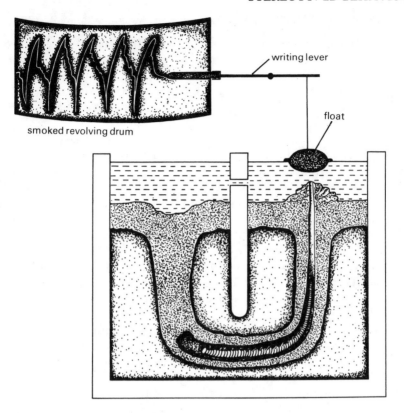

FIG. 69: The lugworm *Arenicola* lives in U-shaped burrows on muddy beaches. In the laboratory it can be made to write information about its activities on a revolving, smoked drum. Alterations in the flow of water through the burrow due to activities such as irrigation, feeding or defecation can be detected using a writing lever attached to a float. (FROM WELLS)

A turning reaction not followed by locomotion can be observed when locusts are 'cold' in the early morning. They orient themselves with their long axes at right angles to the sun's rays, thus presenting a large surface area to gain heat from radiation.

As with reflexes, it is not possible to classify orientation reactions in a wholly satisfactory way. The terms adopted here for classification have been used and found useful over a number of years, especially for descriptions of invertebrate responses. The terms describe the *type of response* an animal shows and should not, as has been done on occasions, be confused with the *mechanisms* which allow their possessor to perform a particular type of reaction.

127

THE BEHAVIOUR OF ANIMALS

PRIMARY ORIENTATIONS

This category of response overlaps with that already described as position reflexes (p. 123). Animals maintain a basic orientation to important stimuli such as gravity, light or contact. Man's basic orientation is erect with only two limbs contacting the ground, but most mammals have four limbs on the ground. Swimming animals usually present their ventral surfaces towards gravity.

Superimposed on this basic position in space are secondary orientations whereby an animal, still maintaining its primary orientation, steers or turns in relation to various stimuli.

Most animals have a bilateral symmetry; if a limb or organ is found on one side of the body, a similar structure is usually located on the opposite side. Such an animal receiving unilateral stimulation usually suffers, via reflex responses, more or sometimes less activity on the side exposed to stimulation. Often the result is a change in the animal's direction or position. A response of this type was formerly called a *tropism*, a term first introduced by botanists to describe the turning movements of plants. Since different mechanisms serve to produce movements in animals and plants, tropism was replaced by *taxis*.

There are a number of different types of taxes, but before describing these we will consider the reactions where no orientation occurs with respect to the source of stimulation. These undirected responses are called *kineses*. With change in stimulus intensity there is an alteration in either the rate of locomotion (*orthokinesis*) or the rate of turning (*klinokinesis*).

ORTHOKINESIS

An increase in the rate of locomotion with increasing stimulation is a common reaction given to many different stimuli, the effect of the response being a removal from adverse conditions to favourable ones. Some ants and caterpillars react in this way in response to temperature increase, and fly maggots react to light intensity increase.

It is suggested that this type of behaviour will increase the chance of locating a particular goal or will retain an animal in a zone of optimal conditions. The fruit fly *Drosophila*, searching for food, flies more rapidly in odourous air. A greater distance per unit time is thus covered, increasing the chances of food location.

The woodlouse *Porcellio scaber* tends to aggregate in dark, moist places due to an orthokinetic response. With decreasing relative humidity speed of locomotion increases, while high humidities result in slower movements or even cessation of activity; the woodlice thus spend more time in such conditions.

KLINOKINESIS

This is a response where the rate of turning depends on the amount of stimulation. When the body louse *Pediculus humanis* is searching for its host, man, it shows klinokinetic responses to odour, temperature and humidity. A louse not detecting any of these 'host near' stimuli moves in a fairly straight path. If a favourable area is entered, the straight path is maintained. But on entering a zone of decreasing stimulation after experiencing favourable stimuli, the louse begins to turn. Keeping a straight path when appropriate stimuli are constant or increasing, and turning only when they decrease, allows a louse to eventually find its host (Fig. 70).

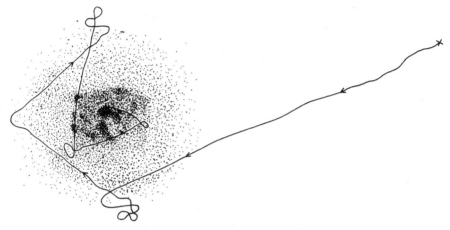

FIG. 70: The human body louse *Pediculus humanis* turns frequently only when the intensity of favourable stimuli decreases. This type of response will allow the louse to eventually locate the source of stimulation. (AFTER WIGGLESWORTH)

Some animals simply turn more in a favourable area, thus spending more time there or, if searching for a goal, increasing their chances of finding it. The small crab *Pinnixa* lives commensally with a tubeworm. If the crab is removed from the tube and subjected to a current of water containing a chemical factor from its partner worm, it turns much more often than in a stream of water which lacks this 'host factor' (Fig. 71).

Ladybird beetle larvae (coccinellids) feed on aphids, and when searching for these on leaves they follow a more or less random path, turning occasionally. At intervals they stop, fix the abdomen to the leaf and swing the anterior part of the body from side to side. This effectively increases the area over which they search. After a coccinellid larva has found and fed on an aphid, its path of movement includes many more turns than previously (Fig. 72); this behaviour keeps the larva in the general area where

129

it has been successful once, and where it has a good chance of further success as aphids usually occur in groups.

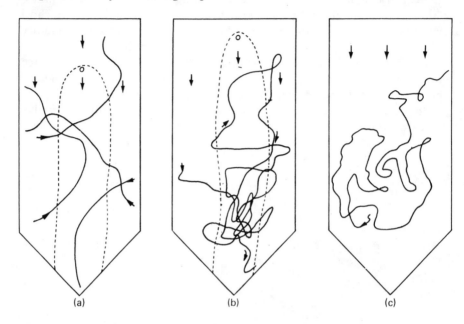

(a)   (b)   (c)

FIG. 71: The crab *Pinnixa chaetopterana*, when removed from the host (tubeworm) and subjected to a current of water containing a chemical factor from its partner worm (host factor), turns much more often than in a stream of water without the host factor. The stream of water enters through a hole indicated near the top of the figures (a) and (b), and occupies the area enclosed by the dotted line. When there is no host factor in the stream (a) the movement is not restricted, but when the host factor is added to the stream (b) or diffused in the whole area (c) the crab turns more frequently and tends to stay within the influence of the host factor. (FROM CARTHY)

Taxes

Animals reacting kinetically are not oriented with respect to the source of stimulation. With *tactic* responses, the animal takes up a particular direction with respect to the stimulus source. Locomotion often accompanies these responses and can thus produce movement towards, away from, or at an angle to the source. The oriented *turn* into a position with respect to the stimulus source and the *maintenance* of this position is the tactic response, not the entire locomotion. Several categories of taxis are recognized.

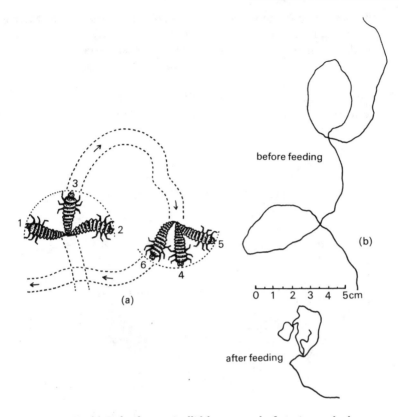

before feeding

(b)

0 1 2 3 4 5cm

after feeding

(a)

FIG. 72: (a) Path of a coccinellid larva on a leaf. At intervals the larva stops, fixes the abdomen and swings from side to side. The area over which it searches for its aphid prey is increased by such movements. (b) After feeding, the larva turns more frequently than before and this behaviour keeps it near the aggregation of aphids. (AFTER BANKS)

KLINOTAXIS

The animal appears to make successive comparisons of the light intensity on either side of the body by pendulum-like movements of the body or its appendages.

Fly maggots just after their last meal before pupation are strongly photonegative. Their photoreceptors are found at the bottom of two pockets, one on either side of the body near the anterior end. In this position, light coming from behind the animal will not stimulate the photoreceptors. Klinotaxis in maggots is best seen after they have been dark-adapted for a period. In a beam of light from a spotlight or torch the maggots move away from the light source along a relatively straight path, but swing the anterior

131

part of the body from side to side as they go. At each swing to the right, light from behind is detected, producing a bend to the left; at each left swing a bend to the right is triggered. The fact that successive comparisons of stimulus intensity are involved can be demonstrated by a simple experiment performed in dim red light. Every time a maggot makes a swing to one side, say the right, an overhead light is switched on. Then the maggot will move in a circle to the left. If left swings only are illuminated, it will move in a circle to the right.

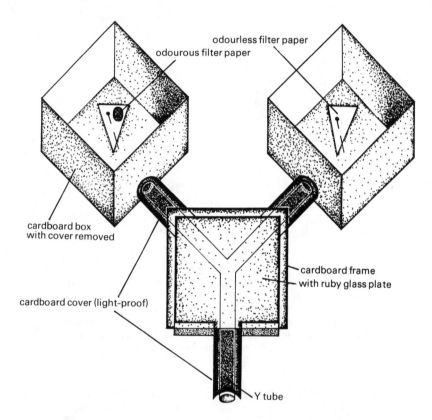

FIG. 73: An experimental design for testing klinotactic orientation towards odours in the honey bee: when one antenna is removed the bee at the choice point of a Y-tube moves the remaining antenna from side to side and detects the arm with odour leading to the food. If this antenna is immobilized the *body* is moved from side to side at the choice point before the bee chooses the odourous pathway. The Y-tube is covered with cardboard except at the choice point where a ruby glass plate is placed to permit observation.
(AFTER LINDAUER AND MARTIN)

Klinotactic responses may be involved in food location by some animals. When planaria (freshwater flatworms) detect juices diffusing from their food they are stimulated into activity. When near the food where the diffusion gradient is marked, the planaria can orient using a klinotactic waving of the anterior parts of the body from side to side.

Honey bees with one antenna removed can locate a food source using klinotactic oscillations. At the choice point of a Y-tube with one arm odour-filled and leading to the food, the remaining antenna is moved from side to side (Fig. 73). If this antenna is fixed in position, side-to-side movements of the whole body occur enabling klinotactic orientation to the goal.

TROPOTAXIS

Many animals can make a simultaneous comparison of the amount of stimulation falling on each side of the body without the side-to-side movements and successive comparisons characteristic of klinotaxis. If this comparison reveals a difference, a turn is made until the animal receives balanced stimulation. This type of reaction has been termed *tropotaxis*.

The woodlouse *Armadillidium* is photo-positive after starvation or dehydration, especially when the temperature increases. A. Müller found that in a two-light experiment most animals moved between the two sources, i.e. along their resultant (Fig. 74a); with unilateral blinding they moved in circles (*circus* movements—Fig. 74b).

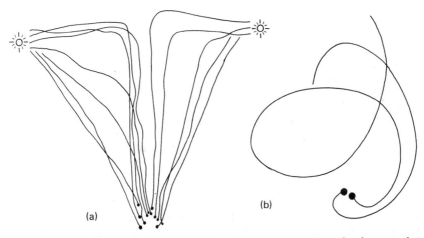

(a)

(b)

FIG. 74: Tropotactic responses of the woodlouse *Armadillidium*: (a) tracks showing photo-positive responses to two lights of equal intensity (AFTER MÜLLER), and (b) tracks showing circus movements of two individuals blinded on the right side, with the light overhead (AFTER HENKE). *Armadillidium* is normally photo-negative, but the sign of the taxis is reversed if the animal is starved or dehydrated.

Some freshwater insects locate their prey by means of a tropotactic response. The backswimmer or waterboatman *Notonecta* locates its prey from vibrations in the water caused by the prey. The long 'rowing legs' are well supplied with sensory hairs and the animal turns until both legs are equally stimulated. It can then swim directly towards its prey. Whirligig (gyrinid) beetles occur on the surface of water, and as they dart rapidly around they can detect vibrations set up by animals caught in the surface film. This they do by contacting the surface film with their antennae. Beetles with their antennae immobilized are unable to orient towards their prey.

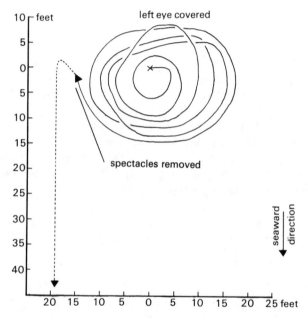

FIG. 75: The green turtle *Chelonia mydas* with one eye blindfolded turned to the side of the uncovered eye, but when the blindfold was removed it gained the correct seaward orientation (AFTER EHRENFELD)

Females of the green turtle *Chelonia mydas* are very short-sighted on land. They lay their eggs at night high up on a sandy beach and afterwards orient towards the sea by means of a tropotaxis. There is more light at the seaward horizon and they use this stimulus to locate the sea. David Ehrenfeld constructed special spectacles with one half a blindfold and placed them on turtles returning to the sea. The turtles turned round and round to the side of the uncovered eye. When the unilateral blindfold was removed, the

134

correct seaward orientation was again possible (Fig. 75). If a filter was placed over one eye, reducing the amount of light transmitted, some of the turtles began circus movements, but eventually they all took up a seaward direction with a deviation towards the side of the uncovered eye.

M. Lindauer and H. Martin introduced honey bees into a Y-tube with odour in only one arm, and found that the bees chose the arm with the odour. If the antennae were crossed and held in this position (Fig. 76), the bees entered the other arm. This tropotactic response was only possible when the antennae were widely separated. When they were fixed with the distal ends less than 2 mm apart, orientation by a tropotaxis was no longer possible, though side-to-side (klinotactic) movements, as described earlier, now made orientation to the goal possible.

K. Herter illuminated the fish louse *Argulus* from below and found that the animal turned on its back to swim. Unilateral blinding caused it to roll and spiral while swimming, indicating that balanced stimulation is necessary for its normal responses.

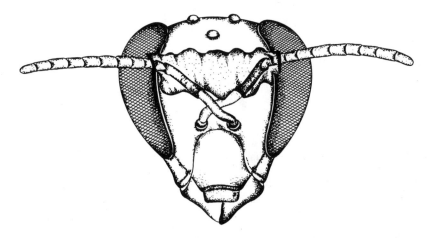

FIG. 76: Head of a honey bee with antennae fixed in a crossed position: in the apparatus shown in Fig. 73 such bees entered the arm without the odour (tropotactic response) while normal bees entered the odour filled arm (AFTER LINDAUER AND MARTIN)

TELOTAXIS

W. von Buddenbrock placed photo-positive hermit crabs near two lights of equal intensity and observed that they oriented towards one light only. Occasionally orientation was changed in favour of the other source, producing a zig-zag track (Fig. 77).

Honey bees in a two-light experiment show similar types of responses.

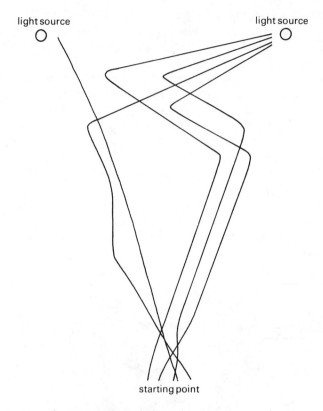

light source   light source

starting point

FIG. 77: Telotactic responses of hermit crabs: when von Budden-brock placed two equal lights near photo-positive hermit crabs, they oriented towards one light only and occasionally changed the course to the other light, forming zig-zag tracks (FROM FRAENKEL AND GUNN)

In another experiment with honey bees, D. E. Minnich blackened their left eyes and allowed them to walk in a beam of light. Some initial turning to the right occurred, but after a number of runs with the same individuals, direct orientation towards the source was possible (Fig. 78).

These reactions of animals are said to be *telotactic*. Unlike klinotaxis and tropotaxis where balanced stimulation is achieved by the animal, telotactic orientation does not demand balance; orientation is possible even if one receptor of a symmetrical pair is removed. In a two-light experiment an animal will go towards one or the other of the sources (assuming a positive response) and not along their resultant, as would animals responding klino- or tropotactically.

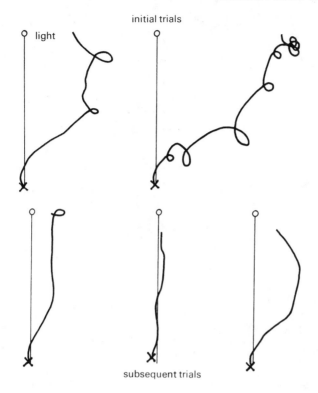

initial trials

light

subsequent trials

FIG. 78: Minnich blackened the left eye of a honey bee and allowed it to walk in a beam of light. Initially the bee showed circus movements to the right but later gradually straightened the course towards the light. (FROM FRAENKEL AND GUNN)

MENOTAXIS

In the types of taxes considered so far, the animal orients either towards or away from the source of stimulation. When an animal orients at a constant angle to the stimulus source, the response is called *menotaxis* or sun compass orientation. Menotaxes are seen in ants and bees which find their way back to a nest or hive. Many species of ants can be induced to change the direction in which they are moving if they are shaded from the sun and the sun's rays are directed on them from a different position by a mirror (Fig. 79).

Foraging ants returning to their nest utilize the beacon provided by the sun to indicate direction. V. Cornetz lifted ants on their way back to the nest and put them down in a different position. They moved off in the same

137

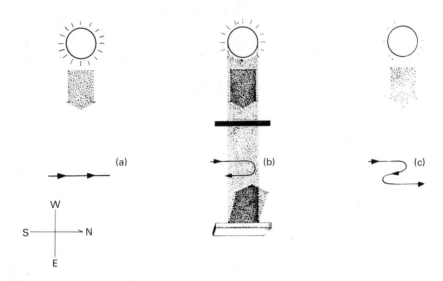

Fig. 79: An ant returning to its nest maintaining the course with the sun on the left (a) was interrupted by Santschi, who shaded the direct rays of the sun with an opaque screen and used a mirror to illuminate the ant from its right side (b). The ant changed direction so that its left side continued to receive illumination, but when the opaque screen and the mirror were removed the ant took up its original direction (c). (MODIFIED FROM FRAENKEL AND GUNN)

compass direction that they were following earlier. If returning ants are kept under a light-tight box for a period, they take up a new angle of direction when the box is lifted. The new path is at an angle to their original direction approximately equal to the angle the sun moved while the ants were covered. Inexperienced foragers may be 'fooled' in this way, but as the ants mature they learn to make allowance for the sun's movements and thus increase the accuracy of their navigation.

Honey bees make extensive use of the sun compass reaction and, like the experienced foraging ants, compensate for the movement the sun makes during their activities. E. Wolf performed an experiment in a disused aerodrome where conspicuous landmarks were lacking, and showed the importance of menotaxis in returning foragers. Bees were captured at a feeding table, quickly transferred to new positions and released. The path they followed was parallel to the one between the hive and feeding table (Fig. 80). The bees not only took up the 'correct' direction towards their hive, but also flew the 'correct' distance before they started random searching movements. It is not vital that the sun be visible to bees at all times for them to navigate successfully. At the sun's position, even through fairly dense cloud cover, there is a stronger penetration of ultraviolet light which can be appreciated by bees. In addition, if a patch of blue sky is present bees can utilize the pattern of polarized light, which is dependent on the position of the sun at that particular instant, for orientation.

feeding place

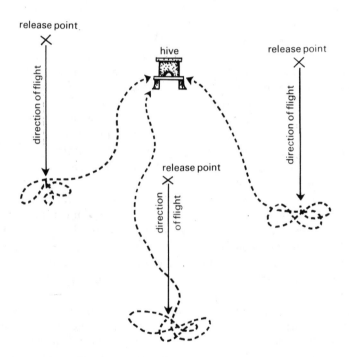

FIG. 80: Wolf experimented with honey bees to demonstrate their sun compass reaction in a disused aerodrome where there were no conspicuous landmarks. The diagram shows the flight paths of honey bees after capture at a feeding place and release at the places indicated. The initial directions flown were parallel to the directions the bees would have flown to return to their hive, and the initial distances flown were approximately equal to the distance from the feeding place to the hive. Bees then showed random searching movements and eventually found the hive. (FROM FRAENKEL AND GUNN)

ROLE OF ORIENTATION REACTIONS

We have seen that orientation reactions are utilized to achieve aggregation in zones of optimal conditions; to maintain the correct position in space; to direct an animal towards a goal; to allow foragers to find their way back to the nest. But we must not be misled into thinking that invertebrate behaviour consists only of simple orientation reactions. Many of these reactions have been studied in the laboratory with one type of stimulus varying, others being controlled or absent. In the field, with many and complex stimuli interacting, the distinctions between orientation responses often become less clear.

The sign of a taxis, either positive or negative, may alter, and this can depend on numerous factors such as temperature, availability of food or water, age of the animal, degree of dark adaptation and many more. The beetle *Blastophagus pinniperda* is photo-positive in spring from 10-35°C, but photo-negative at other temperatures. In autumn it is photo-positive only within the range 20-30°C. Thus the same temperature in spring or autumn, say 15°C, induces positive or negative phototaxis respectively. A small mite *Unionicola* lives inside the shell of the freshwater mussel *Anodonta*. Normally the mites are positively phototactic. But if we investigate their responses in water taken from a pond or creek containing their partner mussels, they show a negative phototaxis. This response, of course, increases the chances of their locating a host.

Bees use the position of the sun or the plane of polarized light in the sky to keep track of the direction of flight. This does not mean to say that they do not use other cues in orientation. They can also pilot a course by using conspicuous landmarks such as trees, rocks or houses. To test the relative importance of the orientation methods known to be used by bees, Karl von Frisch and his associates in Germany designed an experiment. They first trained bees to find a source of food 180 m away to the south of their hive. The path between the hive and food was found on the edge of a forest running from north to south. Then the colony was moved to a new unknown landscape with a similar forest edge running from east to west. The bees looked for the source of food not in the trained direction to the south of the hive, but along the edge of the forest 180 m to the west! Thus the conspicuous landmark (forest edge) was used in orientation in preference to the sun compass. It was found that unbroken landmarks such as a forest edge, shoreline, road, etc. provide a means of orientation superior to compass orientation, but that the latter is superior to isolated landmarks such as scattered tall trees or clusters of bushes.

As the above experiment demonstrates, an animal probably uses more than one type of stimulus in the complete orientation process. The relative importance of different kinds of stimuli differs according to the animal and its environment.

NAVIGATION

Many of the examples of behaviour considered under the heading of taxes, especially menotaxes, are examples of navigation. We do not usually describe a gyrinid beetle moving towards its prey as navigating, but this depends on our definitions. If navigation is taken to mean how animals find their way about, then the majority of tactic responses are navigations. Most definitions of navigation include a mention of long-distance movements and the animal's capacity to establish and maintain reference to a goal without using landmarks.

ODOUR

In many social insects, odour trails are extensively employed by foragers to indicate the route to a food source. The fire ant *Solenopsis* marks the trail from a food source back to the nest by allowing its sting to touch the ground at intervals. A drop of secretion (a pheromone) is deposited each time. The odour trail is relatively ephemeral; the odour from a 20 cm trail evaporates in about 120 seconds. This ensures that old trails leading to worked-out food sources are not followed, and that if a trail having a high odour concentration is encountered, it leads to a source currently being exploited by many workers.

Some primitive stingless bees mark stones, trees or bushes every two or three yards on their way back to the nest. There the scout bee runs around bumping into other workers and exchanging food samples. When a number of workers have been recruited, the scout leads the group back along the scent trail to the food.

Young salmon when they hatch in a stream enter the river and migrate to the sea. After a few years they return to the particular stream of their birth. They do this by retaining the memory of the stream's odour for several years and detecting this after they return to the river from which they entered the sea. However, this cannot account for their ability to find the 'correct' river when they return from the sea. Young salmon tagged in Canada have been recovered after migration only from that river in which they were originally caught.

CELESTIAL CUES

As described earlier, honey bees in the absence of conspicuous landmarks will navigate by reference to the position of the sun. In many well-documented examples of navigation we are ignorant of the cues the animals use to achieve their goal, though celestial cues (the sun or stars) are often certain to be involved.

Green turtles travel from Brazil to Ascension Island in the South Atlantic, covering the high seas for at least a thousand miles to find a target only five miles across. Recoveries of tagged turtles have shown that they scatter over a wide range of latitudes ($2°$ to $20°$S) along the Brazilian coast, and yet all return to this one island for breeding after three or four years. As in salmon, the migratory paths and the method of orientation used by turtles are still unknown.

Australia is a wintering area for two species of swifts and many species of waders that breed in the northern temperate and subarctic regions. These birds arrive in the Australian spring and depart in autumn, except for a few that remain in winter. Mutton birds (shearwaters) show a similar seasonal pattern of movement, but they breed in Australia and winter in the North Pacific. Some of the young short-tailed shearwaters, *Puffinus tenuirostris*, banded while still unable to fly on Griffith Island, Victoria, were recovered about two months later off Hokkaido (Japan) and Sakhalin (U.S.S.R.) after travelling at least 5700 and 6200 miles respectively.

In these cases, young birds migrating for the first time are unaccompanied by parents and often fly over the ocean day and night. We still do not know what causes them to migrate or how they find their way. Does each individual navigate independently while migrating in large numbers? What environmental cues do they use in navigation? They might use celestial cues, but any east-west drift would make it impossible for them to correct the course of migration, unless they could at any instant use the sun or stars to compare the local time with that of their destination.

Two German zoologists, Franz Sauer and his wife, experimented with garden warblers under the artificial sky of a planetarium to test methods of star navigation. Both experienced birds and hand-reared birds with no previous opportunity to see a natural sky behaved in the same way when tested in the planetarium. When no stars were shown in a lighted planetarium chamber, the birds were completely disoriented. Results of the experiments were interpreted in terms of this control test. Under artificial skies the birds tried to fly in the direction of the normal migration route relative to the star configurations. The garden warbler migrates in autumn from Germany to Africa, first flying south-easterly and then changing course to a southerly direction over the eastern end of the Mediterranean. The Sauers claimed that this alteration of directions could be induced by shifting the planetarium sky from the latitude of the breeding ground to a setting of Turkey. However, another German zoologist, H. G. Wallraff,

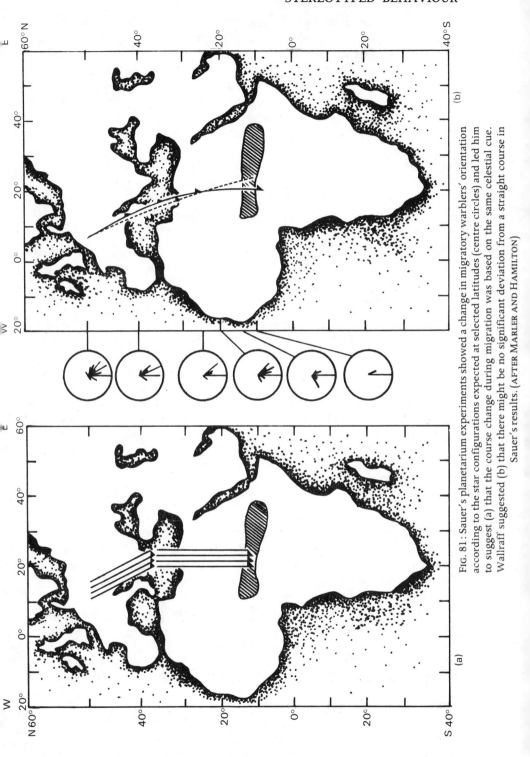

Fig. 81: Sauer's planetarium experiments showed a change in migratory warblers' orientation according to the star configurations expected at selected latitudes (centre circles) and led him to suggest (a) that the course change during migration was based on the same celestial cue. Wallraff suggested (b) that there might be no significant deviation from a straight course in Sauer's results. (AFTER MARLER AND HAMILTON)

challenged the statistical validity of their experiments (see Fig. 81). Further experiments are now required to settle this question.

Along the east coast of Australia unseasonally strong westerly winds take many birds completely off course during their autumn migration. For example, Willis Island, about 250 miles east of Cairns, in the past has received many species of land birds including peewees, martins and pigeons. Migrating mainland silvereyes often drift to islands of the Great Barrier Reef, where they mix with local populations for varying times. Such off-course migration is called *drift migration*. It is not known whether land birds blown so far off course can return home. Some Tasmanian silvereyes were blown across to New Zealand over a century ago and settled there, greatly increasing their numbers. Welcome swallows and spur-winged plovers of Australia have also settled in New Zealand in recent years. On the other hand, white-fronted terns and banded dotterels of New Zealand migrate to Australia regularly in winter, providing examples of rare east-west migration.

Birds and many other animals can certainly determine *directions* from the sun and from the stars, but there is as yet no proof that they can determine their *displacement* from a given locality by celestial cues.

Besides celestial cues there are possibly others that might be used by animals, for example the directional variations of the earth's magnetic field, and variations of forces resulting from the earth's rotation. So far, all attempts to demonstrate the ability of birds to detect these have failed. If you were carried away, blindfolded, to an unknown area and then given a compass to find your way home, could you walk straight home? The answer is, no, unless more information is available. What is this extra information which must be used by homing pigeons and other displaced birds?

CYCLIC BEHAVIOUR

ENDOGENOUS AND EXOGENOUS RHYTHMS

Cyclic activities or rhythms are met with, and shown, throughout the life of all animals. Locomotion is always rhythmic, whether in water, in air or on land. The tails of fish, the wings of birds, the limbs of vertebrates and the appendages of insects all function in a rhythmic way. Practically all the physiological activities of animal life are rhythmic. There is a well-marked rhythm in the activity of the digestive organs, and often in their secretions. The activities of the gall bladder, urinary bladder and uterus are also rhythmic. All muscular tissues possess the power of rhythmical contraction to a greater or lesser degree. The vertebrate heart, in its periodicity, provides a good example of a physiological rhythm, for the rhythmicity of the heart is inherent in its make-up. As soon as they are formed, the microscopic heart cells of a developing embryo beat with an automatic rhythm, long before nerves have reached them or the blood has formed.

Breathing in mammals is partly brought about by the rise and fall of the diaphragm. This rhythm of activity is under nervous control, for if the nerves to the diaphragm are cut, it remains motionless. Centres in the brain are constantly giving off rhythmical stimuli which pass via the nerves to the muscles of the diaphragm, producing its rhythmical contractions and relaxations.

The feeding activities of the lugworm *Arenicola*, described on p. 126, provide an example of a simple physiological rhythm. Feeding is in short bursts of a few minutes duration, with a few minutes rest in between. Rhythmically arising activities in the oesophagus cause the intermittent feeding activity of the intact worm. Under natural conditions the lugworm performs irrigation movements, circulating water through its U-shaped burrow. It also moves to the surface to defecate every twenty to sixty minutes. A fasting worm in a glass tube will perform similar activities even though it has no residues to discharge. The ventral nerve cord controls the irrigation and defecation activities.

Many physiological rhythms are involved in the multitude of activities and functions that an animal performs, and they are closely associated with the successful organization and hour-to-hour economy of the living organism. These rhythms range in period length from those performed thousands of times per second to those of a minute's or an hour's duration. They can be related to the 'private' or 'physiological' time of the organism and are relatively unaffected by the 'outside' or 'astronomical' time. There are other rhythms in animals which have a longer period and which parallel or are induced by rhythmic factors in the external environment. These can be correlated directly or indirectly with the action of the sun, earth or moon, and are tidal, daily, semilunar, lunar (monthly) or annual cycles.

The solar day is associated with a cyclic change of illumination, temperature, humidity and barometric pressure, all of which are important to life and are used by organisms to regulate their activities. Another natural period of importance to animals and plants is the lunar day. The influence of the moon is felt most in the intertidal regions of the shore, because the tides are produced predominantly by the gravitational pull of the moon. Hence the tides are of a lunar frequency and occur about fifty minutes later each day. At approximately fifteen-day intervals, the influence of the sun and moon is additive and produces the high 'spring' tides.

Most animals possess a daily activity pattern of one kind or another. Some animals, including butterflies and sparrows, are *diurnally* active; others are *nocturnal* like cats, bats and earthworms; still others, such as the death

BIOLOGICAL RHYTHMS

Frequency or period length of cycle (approximate values)	Type of activity
SHORT-PERIOD 'PRIVATE' RHYTHMS	
1000 per sec or more	Wing-beat frequency in midges
200 per sec	Wing-beat frequency in honey bees
1000 per min	Canary heartbeat rate
70 per min	Human heart rate
18-20 per min	Human respiration rate
Every 6-7 min	Feeding activities in the lugworm *Arenicola*
Every 20-60 min	Irrigation and defecation activities in *Arenicola*
LONG-PERIOD 'EXTERNAL' RHYTHMS	
12.4 hr (tidal)	Opening of shell valves of the oyster *Crassostrea* Locomotory activity in the fiddler crab *Uca*
24 hr	Emergence of adults of *Drosophila* flies from pupae Nocturnal activity of many mammals
15 days	Activity in the periwinkle *Littorina rudis*, which lives high up the beach and is submerged every 15 days on spring tides Spawning in the grunion fish *Leuresthes*
29.5 days (lunar cycle)	Reproduction in the Bermudan fireworm *Odontosyllis*
Annual	Reproduction in many birds and mammals

adder and the bat-hawk, are active during dusk and dawn (*crepuscular*). Many animals that live in the intertidal regions of the shore have behaviour patterns that are repeated with the tides, each cycle averaging about twelve and a half hours in length, the duration of the intertidal period. Certain crabs feed at the water's edge only on ebb tides, while oysters and clams feed actively at high tides. Some intertidal animals, especially those that live high up on the beach so that they are only submerged at the high spring tides, have approximately fifteen-day periods of activity.

It is of vital importance to animals with a breeding season that all the males and females should synchronize the development of gonads. For terrestrial animals, in which fertilization is internal, both male and female sexes should be fit for intercourse at the same time. This is brought about by a reproductive rhythm which may be correlated with the seasons. Liberation of the gametes of marine organisms into the sea should be synchronized so that the probability of them fusing with each other will be high.

During full moon of every summer month, the Atlantic fireworm swarms in the waters off Bermuda. The worms, luminescing brightly, swarm fifty-five minutes after sunset. Breeding takes place, and only stragglers are left after about half an hour. In Fiji and Samoa the Palolo worm lives in crevices in the coral rock. Once a year the posterior parts of the male and female worms break off and rise to the surface of the sea to release the sex cells. The local natives regard these worms as a delicacy and know precisely when to make the trip to collect them, for the swarming occurs before dawn on one or two days only, at the third quarter of the October or November moon. The worms are collected by the ton, and may be eaten raw, boiled or fried.

These examples indicate some of the ways in which animals time certain activities with great precision. Two questions now arise: are these activities set off as a direct response to rhythmic factors in the external environment, or do these animals possess an innate clock or calendar which controls these activities and thus causes them to be set off at the right time independently of external stimuli? The answer is yes to both questions, for some animals behave in the first way, others in the second.

When such a periodicity is triggered directly by environmental factors it is termed *exogenous*. The carnivorous marine snail *Nassarius festivus* is active only during darkness. S. Ohba reversed the natural periods of light and darkness by illuminating the animals at night and keeping them in darkness during the day. The snails reversed their activity periods. It was also possible to impose on them a rhythm with a different period length; for example, an eight-hour cycle of activity and inactivity was induced by giving eight hours darkness followed by eight hours of light. Such an imposed rhythm could be reduced to six or even two-hour cycles. Animals possessing exogenous periodicities can sometimes be induced by rare natural events to perform certain activities at inappropriate times: during a total eclipse, midges may swarm as the light intensity falls, tree frogs may croak and some domestic fowls start crowing.

However, many rhythmically performed activities are *endogenous*; they are not triggered by external events and when the animal is maintained in constant conditions the rhythm still persists. Twice in every twenty-four hours the activities of many intertidal animals change, and this rhythm is retained even if these animals are kept under constant dim illumination in tideless aquaria. Similarly, many twenty-four hour activity patterns in animals continue to run with a more or less twenty-four hour period when the subjects are kept in constant light or darkness at constant temperature. Such persistent rhythms are widespread in both the animal and the plant kingdoms.

EXAMPLES OF ENDOGENOUS RHYTHMS

- Bioluminescence in the flagellate protozoan *Gonyaulax*
- Oxygen consumption in potatoes and crabs
- 'Sleep' movements in the leaves of bean seedlings *Phaseolus*
- Pigment migration in the eyes of the isopod crustacean *Ligia*
- Photo-positive responses in the green protozoan *Euglena*
- Colour change in the fiddler crab *Uca*
- Egg release in the brown alga *Dictyota*
- Potassium, sodium and chloride excretion in humans

Animals showing such rhythms are said to contain a *biological clock*, and they use the clock in many ways: to control the twenty-four hour colour change suffered by the fiddler crab *Uca*, to time the onset of nocturnal running activities in the cockroach, to time the early-morning emergence of *Drosophila* flies from pupae so that the desiccating midday sun is avoided, and to allow time compensation in those animals which navigate by reference to the sun or moon.

BIOLOGICAL CLOCKS

The rate of most of the body's physiological reactions increases by a factor of two or more for every 10°C rise in temperature. But biological clocks have an important property; they are more or less independent of temperature, and have the same period length over a range of temperatures. The rhythm of colour change in the fiddler crab (dark-coloured during the day and light-coloured during night) is almost the same when the animals are kept at 26, 16 or 6°C. If bees kept at 23°C are trained to come to a feeding table at a certain time of day, they still come at the same time when kept at 31°C (or perhaps a little early—but bees have a habit of coming early to a feeding table in training experiments of this kind). The value of temperature-independent clocks is obvious—they would be of little use if they suffered a speeding-up or slowing-down every time the temperature rose or fell.

Although a rhythm can persist under conditions of continuous light or continuous darkness, this is not to say that light has no effect on the rhythm. Cockroaches become active at dusk; their activity reaches a peak soon after dark and declines during the remainder of the dark period and the following light period. If the normal light and dark periods are reversed and maintained, the running activity begins at a slightly different time each day until the rhythm is in phase with the new conditions. The cockroach has reset its clock.

Under natural conditions, animals do not experience such drastic changes in the light/dark cycle; an animal would have to be transported to the opposite side of the world to be exposed to complete reversal of light and dark periods. But animals are exposed to slight changes in daylight length (and night length) throughout the seasons, and are able to keep their rhythmic activities in phase with the external conditions by slight clock resettings.

In some animals, the site of a biological clock is known. In the lugworm, feeding is triggered by the activity of the oesophagus (p. 126). The cockroach *Periplaneta americana* will lose its running rhythm if it is kept for a few weeks in continuous bright light. Janet Harker, working at Cambridge, took such an arhythmic cockroach and joined it back-to-back with a roach having a normal rhythm of activity (Fig. 82). This allowed the exchange of body fluids between the cockroach pair, but there was no nervous connection. The lower cockroach was the arhythmic member of the pair. In continuous light the two animals showed a rhythm of activity similar to that previously shown by the upper immobilized roach. Because only body fluids could be exchanged between the two animals, it was concluded that some factor in the blood of the upper cockroach passed to the lower one and induced an activity rhythm.

Cockroaches will carry on moving quite actively for up to ten days after their heads have been removed. These headless cockroaches, however, have no activity rhythm. If the suboesophageal ganglia are removed from a normal rhythmic cockroach and transplanted into a headless roach, an activity rhythm will be induced. It is known that these ganglia produce a secretion, and it is this secretion that influences the behaviour pattern of the animal. The ganglia are influenced, via sense organs, by illumination changes.

These suboesophageal ganglia can be chilled *in situ*, and when this is

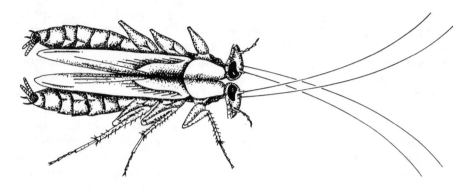

FIG. 82: Harker joined two cockroaches *Periplaneta americana* in parabiosis to investigate the effect of body fluids on the control of rhythmic activity. The lower arhythmic roach became rhythmic after receiving body fluids from the upper rhythmic roach.

done there is no alteration in the running activity of the cockroach. If the chilled gland is transplanted into an arhythmic cockroach, the recipient roach shows a rhythm of activity not identical with that of the donor, but the rhythm is delayed by the number of hours equal to the period for which the gland was chilled. This indicates that the suboesophageal clock is subordinate to some other 'higher' clock.

It is quite probable that every cell of the body has its own biological clock. Idaho potatoes have a twenty-four hour rhythm of oxygen consumption, and each cell is not very different from all other cells. Cell division and light production in the flagellate *Gonyaulax* and mating reactions in *Paramecium* are endogenous activities. Since protozoans are unicellular, the clock in these animals is located within the single cell.

The existence of biological clocks is significant not only for the animals which possess them, allowing the precise timing of various activities, but also for students of animal behaviour. We should be aware that some activities are performed much better at a certain time of day, and that some activities are only performed at appropriate times. If we run earthworms in T-mazes to investigate their learning ability, performance will be much better during night runs than during day runs. The spiny lobster *Jasus lalandai* and the swimming crab *Portunus pelagicus* are nocturnal animals. If a study of their feeding activities is planned, the student should be prepared to work during dusk or at night, because food offered during daylight is usually ignored. Of course, an astute student could always reset the lobster or crab clocks with an appropriate light/darkness cycle, and induce activity at the desired time of day to avoid night work.

A word of caution is necessary with regard to the slightly frivolous suggestion offered above. Some rhythms, like crab activity, are fairly easily reset with illumination changes. Others, such as the rhythm of potassium excretion in humans, are not easily modified. The rhythms of the many different organs and organ-systems in our bodies are normally phased to allow their efficient working and co-operation. One reason why we may not 'feel right' for some time after a long plane journey is that our biological clocks, all resetting to the new time zone, are not resetting at the same rates. It would be advantageous if astronauts on extended flights were not held in constant conditions, but were given external stimuli such as a light and dark cycle from time to time. This would allow their numerous biological clocks to resynchronize with each other.

The importance of biological clocks and their synchronization was demonstrated by Harker during her investigations into the running rhythm of cockroaches. Cockroaches were held in reversed periods of light and darkness until their rhythm of activity followed the new light and dark régime. The suboesophageal ganglia from these roaches were implanted into roaches with a normal rhythm. The experimental animals now had two pairs of suboesophageal ganglia, but they were twelve hours out of phase with each other. All of these insects died from intestinal cancer! In a control group where the rhythms of both donor and recipient were the same, there were no such deaths.

Animals make use of their biological clocks in many ways, some of which have been mentioned above. An important function is that of pre-adaptation; the animal is 'alerted and ready' for activity with the forth-coming change in conditions. If honey bees are trained to feed at a certain time of day, they usually appear a few minutes early on subsequent days. They appear 'aware' of the imminent supply of food and are ready to exploit it. They do this under natural conditions, for some flowers offer pollen and nectar only at certain times. In this case the clock can reduce the incidence of unrewarded flower visitation. An intertidal animal which feeds at a certain time, say on low tides, can utilize the whole of this limited time period if it is made primed and ready by using its clock. The alternative is to wait for some external trigger, say desiccation, to initiate activity.

COMPLEX INNATE BEHAVIOUR

Stereotyped behaviour is not restricted to the relatively simple activities so far described, but also occurs in more complex instinctive behaviour. As mentioned in the second chapter, ethologists discovered many complex behaviour patterns consisting of fixed action patterns. These action patterns are illustrated by displays of threat or courtship where behaviour patterns

151

are basically innate (not acquired through experience). The triggering of the fixed action pattern is governed by the internal state or *drive* of the animal and the appearance of the appropriate external stimulus, the *releaser*. Once triggered, the action patterns do not require the continued presence of the releaser. Under natural conditions such innate behaviour patterns are generally adaptive, tending to ensure survival and successful reproduction.

Ethologists emphasize the above features as characteristics of instinctive behaviour, whereas psychologists approach the subject by analyzing the stereotyped behaviour that results from specified conditions of hunger, thirst, rage, fear, sleep or sexual drive.

Ethologists at Oxford led by Niko Tinbergen investigated instinctive behaviour of a small freshwater fish, the three-spined stickleback. Under hormonal control, males of the species establish territories by selecting sites and defending them from invaders (see p. 59). A male attacking an intruder will dart towards its opponent with opened mouth and raised dorsal spines. Sometimes, instead of attacking, it will point its head down and make jerky vertical movements similar to those used when digging in sand. This innate behaviour must be interpreted by the intruder as aggressive behaviour, because it responds to this sand-digging by either attacking or fleeing.

In normal sand-digging the male makes a small depression at the nest-site by boring its snout into the sand. The shallow pit thus made is packed with green algae, which are pressed down and pasted with a sticky substance excreted from the kidneys. The male then wriggles through this mass of algae to form a tunnel.

Having built a nest in its territory, the male makes a series of jerky movements in its search for possible stimuli. As it is brilliantly coloured in the breeding condition its movements make it even more conspicuous. This phase of behaviour is called *appetitive behaviour*, during which a series of instinctive behaviour patterns is interrupted through want of appropriate stimuli (see Fig. 9, p. 20). When animals are in this phase their behaviour is influenced by previous experience. A hungry hawk, for example, will look for prey in the areas where it had successful hunting in the past. The male stickleback may go to the edge of the territory where it previously sighted an intruder. The fish is prepared to release fixed action patterns depending on the particular drive activated; for example, the appearance of another male releases aggression, whereas the appearance of a mature female releases courtship behaviour.

The initial courtship of the male consists of zig-zag leaps towards and away from the female. If the female is not sexually mature she will not respond to the zig-zag dance, but if she is ready to spawn, her response will be to assume an upright posture. The subsequent behaviour patterns

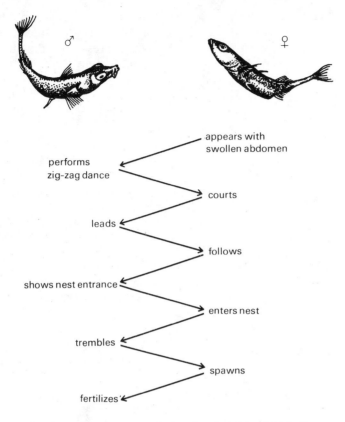

FIG. 83: Courtship sequence of the three-spined stickleback *Gasterosteus aculeatus,* illustrating innate social responses to the partner's sign stimuli (AFTER TINBERGEN)

appear as a series of responses, each being triggered by the partner's action (Fig. 83).

What is the nature of the stimulus that releases such specific responses? Tinbergen used various stimuli, such as models of fish and other objects, to study this question. Results of such experiments showed that the red underside of an intruding fish (typical of male breeding colouration) released the aggressive behaviour of the territory holder. Models lacking red bellies did not invite attack, but as long as the models had the red underside they did not even require the shape of a fish to cause an attack from the territory holder. The swollen abdomen of a mature female, on the other hand, released the initial courtship reaction of the male.

Another experiment of this type conducted by Tinbergen revealed

153

effective stimuli for the begging responses of the herring gull chick *Larus argentatus*. Begging response normally appears when the chick first sees its parent, but the response may be released by presenting the chick with a flat cardboard model of the parent's head and yellow beak, with a red spot near the tip of the lower mandible (Fig. 84). When models with and without the red spot were presented in turn to a number of newly-hatched chicks, the average number of responses to the model lacking the spot was only a quarter of that to the normal model. Models with a spot of abnormal colour released intermediate numbers of responses, decreasing in the order of black, blue and white. When beaks were painted uniformly with various colours, red released twice as many responses as any other colour (including yellow, which was the natural colour of the parent's beak). Models with different shapes of head and beak produced results indicating that the

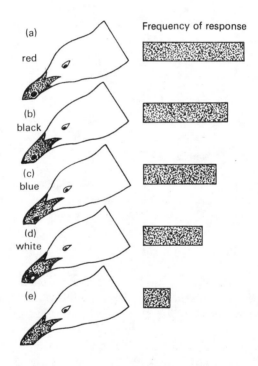

FIG. 84: Tinbergen and Perdeck used cardboard models of heads of the herring gull *Larus argentatus* to demonstrate the effective stimuli for begging responses in chicks. The different colours of the spot on the lower mandible (a–d) and the absence of the spot (e) gave different frequencies of responses in chicks. (FROM TINBERGEN)

shape of the head did not matter; even the beak without the head produced just as many responses. However, to release the begging response the beak had to be thin, elongate and pointed downwards as low and as near the chick as possible.

In natural conditions the parent walks up to the chick and presents its beak downwards, almost vertically, with the red tip in front of the chick. The chick has thus evolved a behavioural mechanism to respond instinctively to symbolized components (sign stimuli) of the environment. If we construct a model in which such sign stimuli are exaggerated relative to the natural object, e.g. a red knitting needle with three white bands near the tip to represent the parent's beak, the model will act as a *super-normal* stimulus and cause the chick to release stronger responses than those produced by the natural object. A person's cosmetic make-up is generally considered to provide such stimuli.

You may or may not eat, depending on the strength of your hunger drive (or other drives countering it) and the attractiveness of the external stimulus (food). Not all innate behaviour is characterized by such a simple all-or-none response. There are instances in which different intensities of a drive or stimulus produce different responses. For example, mild threat of silvereyes consists of a sleeked posture of the body, but as aggressiveness increases, other motor patterns such as wing-fluttering and threat calls are added (Fig. 85). Similarly, the bearded lizard (Fig. 86) puffs up its whole body as threat intensifies.

There are also other cases where sign stimuli can produce variable behaviour. What would happen, for example, if two equally strong stimuli

FIG. 85: Threat postures of silvereyes, illustrating changes of motor patterns with increased aggressive tendency: (a) horizontal posture with sleeked body—weak aggression, and (b) horizontal posture with wing fluttering and threat call—strong aggression (FROM KIKKAWA)

FIG. 86: Threat display of a bearded lizard *Amphibolurus barbatus*

demanded simultaneous release of two conflicting action patterns? Tinbergen used a herring gull to demonstrate behaviour in such a situation. The gull has a tendency to remove red objects from the nest and a tendency to incubate rounded objects. He placed a red egg in a gull's nest and watched the bird's reaction. The gull tried to peck at the egg and also tried to sit on it; the behaviour fluctuated from one response to another. Such behaviour is termed *ambivalent behaviour* and is considered to occur as a result of *competition* between conflicting action patterns in motivation.

Sometimes conflicting situations produce behaviour which does not appear to be related to any of the stimuli or the drives activated at the time. We have seen that a territory-holding stickleback assumes the vertical position of sand-digging when another male intrudes upon its territory. Of course, the intruder does not stimulate the territory owner to build a nest. Similarly, finches wipe their beaks on a perch after feeding, but they may also beak-wipe during courtship when there is nothing to be removed from the beak. Beak-wiping activity in courtship behaviour seems to be displaced from the instinct of maintenance to the instinct of courtship. Such activities are called *displacement activities*. The displacement activity is also known as the *allochthonous* activity, as opposed to the *autochthonous* activity from which it is derived. The autochthonous activity results from the animal's proper drive and serves its original function. As behaviour evolves, the same action patterns are used in different contexts and are often modified when they assume independent functions in displacement activities. As a result it is not always easy to identify autochthonous and allochthonous elements in instinctive behaviour. The courtship sequence of animals, such as that of the three-spined stickleback, may be demonstrated to be composed of a series of displacement activities.

The nervous mechanisms for the elicitation of instinctive behaviour have been postulated by several workers. Tinbergen's original model shows a hierarchical organization of drives. When a certain drive is activated, the behaviour to be expected can be of many different types, but all are limited to movements belonging to that drive. Which type is shown depends on which subdrive, or even which part of the subdrive, is activated. A male stickleback in appetitive behaviour is thus kept ready to react to different situations with different responses under the control of a major drive. In displacement activities the proper drive is considered to be denied opportunity for discharge through its own consummatory act; instead it 'sparks over' to another drive which thus releases behaviour irrelevant to the situation.

The hierarchical organization of nervous mechanisms is supported by some evidence found in the CNS where co-ordination of behaviour is achieved, but there is still a gap in knowledge between the neuro-anatomical features and the supposed pathways of nerve impulses (see p. 115). The concept of the innate releasing mechanism (p. 20) has been developed to explain the selection of stimuli for the release of the particular response. These mechanisms symbolize the specific relations between a particular stimulus character and a particular response. However, as displacement activities show, particular stimulus characters are not necessarily effective for one response only, but may also evoke other responses.

If we call all behaviour patterns that appear in conflict situations displacement activities, then there seems to be an order of priority in which they occur. Some patterns seem to have a very low threshold for appearance (conflict situations do not have to be intense before they appear). For example, preening which occurs frequently in birds is inhibited when birds are engaged in other activities. In conflict situations, however, preening seems to appear quite often. One interpretation is that preening as a result of peripheral stimulation is inhibited while the bird is engaged in other activities, such as incubation or escape from predators, but it will be *disinhibited* if these other activities are suppressed in conflict situations. This hypothesis is called the *disinhibition hypothesis*.

EVOLUTION OF COMPLEX INNATE BEHAVIOUR

As we have seen, not all components of innate behaviour are rigid and unmodifiable. Variations occur while animals are searching for stimuli, and experience modifies the details of behaviour. Of course, not all instincts are apparent when animals are born. In fact, many of the innate behaviour patterns required for survival and reproduction are not functional until they are 'matured'. The pecking response of young birds, for example, is innate and occurs while the young are still being fed by their parents. They cannot eat by themselves until they grow. The behaviour patterns of flight are also innate, but the young do not fly until they are fully grown. This is because the instinct is not matured, and not because the external stimuli that later produce the response are lacking. When ready to fly, they take off and can fly, however clumsy they may look. With each experience they learn to fly more skilfully until elaborate landings in difficult wind conditions become possible.

Moreover, many instincts mature in a particular order. Animals, like most of us, cannot do more than one thing at a time and they are only capable of doing it when internally motivated. Therefore, the maturing of instincts means that early motivation is weak or imperfect. *Play* sometimes functions as an outlet for early motivation and facilitates correct responses in future. Animals often learn vital strategies of survival during play.

Most perching birds spend about a fortnight in the nest, during which time they develop fear responses to moving objects, other than the parents, approaching the nest. Until about ten days old, their response is to crouch flat and motionless in the nest. After this period the same external stimulus causes them to fly out of the nest. This premature leaving is as much an innate response as crouching. In this case the response to the same stimulus changed suddenly as a result of maturing. The adaptive significance of this behaviour is not hard to appreciate if we think of the young bird's chances of survival. While nestlings are still small, to leave the nest means certain death. Although they may all be killed in the nest, there is a slight chance that the predator will not take all of them or that the parents may return in time to save one or two. On the other hand, when they are old enough to fly a few feet at a time, the chance of survival in similar circumstances is greater outside the nest than inside. When the danger has passed, the parents will return to call them from their hiding places and lead them to a new roosting site. Admittedly, life is not easy for the young that have left the nest prematurely as they must suffer increased hazards, but their chances of survival would still be better than if they had stayed in the nest at the time of attack.

After leaving the nest, young birds stop begging for food sooner if they are fed infrequently and given a chance to peck at food themselves. If they are fed frequently they are never strongly motivated to feed themselves. Feeding requires certain skill, which has to be learned, and training is necessary before they can completely drop their earlier instinctive response of begging. Thus in social behaviour, the change of innate behaviour is not as sudden as the change of the fear response. This is because interaction with another individual is important for survival and successful reproduction. If the timetable were rigid, both individuals would have to have an identical schedule. Thus it is easy to see the advantage of not having a rigid timetable of change-over of instinctive acts when maturation depends on environmental conditions and parental responses.

Let us now consider the origin of complex innate behaviour. As has been pointed out, animals can do only one thing at a time. They are, at any given time, engaged in one activity which involves only a few different movements. For example, they may be eating, drinking, or chasing other animals from a territory, but there are not many ways of eating, drinking or chasing. They use certain basic motor patterns which may be varied in different situations. Even a monkey has only about 200 behaviour patterns, including significant variations. Such a list is called the *ethogram* of a species. In an ethogram the instinctive actions that produce instinctive reactions in other individuals (signal movements) are not many, and are often identical with those of other innate behaviour patterns. Beak-wiping used in courtship, for example, seems irrelevant when the beak does not need cleaning. Similarly, why should a puzzled man scratch his head, or a depressed girl unconsciously go to a dressing table? These displacement activities are often derived from instinctive toileting behaviour (care of body surface). They may have had their origin as by-products of nervous mechanisms, and social responses may have evolved making use of such by-products.

When they form part of a display, the movements are very much exaggerated and often rigid and brief compared with the corresponding autochthonous activities. In other words, these activities are *ritualized* through evolution. Ritualization is adaptive as it emphasizes and utilizes sign stimuli which are made simple and conspicuous. It often accompanies morphological changes which are incorporated in the display for maximum effects. The displacement preening of male ducks in courtship, for example, is stereotyped according to the species and very rigid compared with real preening. The mandarin drake (Fig. 87) touches the vane of one of the secondary wing feathers with its beak. The feather structure here has evolved into a huge orange 'flag' and the behaviour draws attention to this structure.

Ritualization also occurs in behaviour which shows 'intentions' of the performer. When a motivation is being built up, animals show *incipient*

FIG. 87: The mandarin drake *Aix galericulata* in its natural position,
showing the 'flag' structure in secondary wing feathers

(*intention*) *movements* which act as releasers to their companions. Co-ordinated sudden movements of flocks of birds or herds of mammals are made possible by such communications. In threat displays an intention of attack appears as a ritualized form (see Fig. 85, p. 155) and acts as a signal movement. Elaborate displays of birds of paradise consist of motor patterns suggesting ritualized intention movements (e.g. wing-arching) and displacement activities (e.g. beak-wiping).

Although behaviour components may be inherited genetically, the evolution of man's social behaviour is almost entirely cultural. Of course, *cultural evolution* is not entirely man's edifice. A local population of muskrats in America, for example, learned to store Indian corn in winter and thereby survived food shortages through several winters. But after one particularly severe drought, most of the local population was wiped out. The next generation did not know about corn storage and could not make use of it. Consequently they suffered greater mortality through the following winter.

Similarly, Japanese behaviourists have recorded instances of potato-washing by monkeys in a wild population. It spread after one monkey developed this habit or possibly copied it from man. If the advantage of such habits is persistent, any genetic change of behaviour patterns in this direction will be favoured by natural selection, with the result that the character becomes fixed as a genetic property of the population. Thus it is not wise to limit the study of innate social behaviour entirely to genetic aspects.

7 Learning

We have discussed the types of behaviour which, on the whole, are not modified through experience. Let us now examine the types of behaviour which emerge as a result of experience or which change through experience. If animals change their responses as a result of experience, we say that *learning* has occurred.

HABITUATION

If you felt an earthquake you would show a freezing or startle response followed by uneasiness. But if earthquakes occurred frequently without causing damage, you would lose such a response, just as you do not rush out to investigate every time you hear a jet plane flying overhead (unless of course you are particularly interested in aeroplanes). In these cases the initial response to an unusual but simple stimulus has disappeared as a result of repeated stimulation which produced no significant consequence to life. This form of learning is called *habituation*. It is different from wilful inhibition or a temporary failure to respond as a result of injury or fatiguing of the nervous system. Habituation is long-lasting and is extremely valuable. If we responded to all the *meaningless* stimuli which surrounded us, our ability to respond to all the *meaningful* stimuli would be seriously impaired.

Animals habituate to many simple stimuli and, as a result, often shape responses which are critical for survival. For example, several species of birds respond to the silhouette of a hawk moving overhead. This response had long been considered innate until a German behaviourist, W. Schleidt, made investigations using the models shown in Fig. 88. He found that turkeys exposed to circles or rectangles overhead were alarmed at first, but became habituated later. These birds, if occasionally exposed to a hawk silhouette, gave strong escape responses. He also discovered, however, that if birds were exposed frequently to a hawk silhouette they habituated to it, while they responded strongly to an occasionally presented circle! Schleidt concluded that the response to the hawk silhouette in wild birds results from their habituation to other types of birds which frequently fly overhead, and it is not an innate response produced only by the hawk silhouette. Would their response, then, differ if hawks were abundant in the neighbourhood?

FIG. 88: Some birds show escape responses to a hawk silhouette (a) moving overhead. Earlier experiments had shown that if the model was flown backwards, therefore looking like a goose, no escape response appeared. Schleidt used this and other models (b and c) to re-investigate such responses. (AFTER SCHLEIDT)

The tubeworm *Galeolaria* lives in the lower intertidal region of the shore, with its calcareous tube firmly cemented either to a rock or to the tube of another worm. It feeds by filtering fine food particles from the water with its crown of tentacles. If there is a sudden decrease in light intensity, or vibration through the water, or if the tentacles are touched, the worm rapidly withdraws into its tube. This withdrawal response is important, for such stimuli could mean the approach of a predator. However, if the worm was living in a region which was turbulent and intermittently shaded, its withdrawal response could be a liability because there would be little time left for feeding if the worm was continually responding. The worm under such conditions would be habituated to changes of light intensity and wave action (see p. 77).

The cockroach will run rapidly if a puff of air is directed at the sensitive cerci at the hind end of the abdomen (Plate II). But if we continually blow puffs of air at the cerci, the cockroach no longer shows this 'innate' escape reaction. A cockroach searching for food under windy conditions would continually give an escape reaction were it not for habituation. The site of this 'learning' is known in the cockroach. Roeder found that after a number of impulses had been transmitted over synapses in certain ganglia the synapses ceased to allow further transmissions. The sites involved were:

(a) the last abdominal ganglia where cercal nerves synapsed with giant nerve fibres of the ventral cord, and (b) the thoracic ganglia where the giant fibres synapsed with the motor nerves to the leg muscles. These latter synapses were particularly unstable (Fig. 89). Information from the peripheral sense organs is *filtered* at various levels of nervous organization; thus it is possible to drop a response by preventing the stimuli from reaching the CNS or relevant effectors.

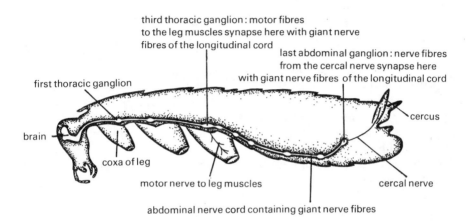

third thoracic ganglion: motor fibres
to the leg muscles synapse here with giant nerve
fibres of the longitudinal cord

last abdominal ganglion: nerve fibres
from the cercal nerve synapse here
with giant nerve fibres of the longitudinal cord

first thoracic ganglion

cercus

brain

coxa of leg

motor nerve to leg muscles

cercal nerve

abdominal nerve cord containing giant nerve fibres

FIG. 89: A cockroach in longitudinal section to show some of the neural elements involved in the startle response: in this neural pathway the synapses in the last abdominal ganglion and particularly the third thoracic ganglion are important in the dropping of the startle response after repeated stimulation. These synapses cease to allow the passage of impulses after a number have been transmitted; thus the startle response becomes inactivated.

In higher animals, however, habituation is considered to involve changes in the CNS. For example, if the intensity of the stimulus to which an animal is habituated is lowered suddenly, the animal will once again produce the response. Karl Pribram cites one interesting case from the police record of New York city. People living in apartments along Third Avenue called the police to complain about strange occurrences that woke them periodically during the night. Investigations revealed that the times of such calls coincided with the times when trains used to rumble past near the apartments on an elevated railway line which had just been torn down. People had previously been habituated to the fearful noise of the train, and the strange incidences that they could not define were the 'deafening silences' that had replaced the expected noise. We may say that *dishabituation* was the cause of their waking.

CONDITIONING

Habituation is a form of learning by which a response is dropped when it is no longer useful or relevant. In contrast, *conditioning* produces a response to a new stimulus by associating the new stimulus with an old one. W. H. Thorpe defines conditioning as 'the process of acquisition by an animal of the capacity to respond to a given stimulus with the reflex reaction proper to another stimulus when the two stimuli are applied concurrently for a number of times'.

The blowfly *Cyanomyia cadaverina* extends its mouthparts when sugar solution stimulates the tarsus. H. Frings introduced the odour of coumarin, an aromatic, to condition the flies. When first tested without sugar solution most flies did not respond to coumarin, but after three periods of training, in each of which the flies were exposed to a combination of sugar solution and coumarin six times, 90 per cent of the flies responded to coumarin alone by extending the mouthparts.

In this case the new stimulus was forced on the fly and the fly did not find it voluntarily. This type of conditioning was first described by Pavlov (see p. 8) and is called *classical conditioning*.

If, instead of sounding a bell and blowing meat powder to the dog as Pavlov did, one gave the meat powder every time the dog scratched the floor, it would eventually associate its scratching of the floor with food. The dog's discovery would then be accidental, but subsequently the association would be established between its own activity and the meat powder (unconditioned stimulus). In nature, animals often take such an *active* part in finding new stimuli. In appetitive behaviour, for example, animals constantly discover new stimuli, and an association is often established, rightly or wrongly, with the old stimulus. Such association established while animals are actively searching for stimuli is not easily lost, even when they are only occasionally rewarded. This type of learning is called *instrumental* or *trial-and-error* learning. In instrumental learning a given response can be reinforced by a variety of rewards and a given reward can reinforce a variety of responses.

In the past the involuntary or visceral responses were thought to be modifiable only by classical conditioning, but Neal Miller and Leo DiCara of Rockefeller University have recently demonstrated that instrumental learning is also possible in the autonomic nervous system which produces the involuntary and visceral responses.

They paralyzed rats with curare so that the animals were completely

164

denied opportunity to move any of the skeletal muscles, while maintaining consciousness and autonomic responses. Such curarized animals cannot even breathe or eat, and therefore there was no possibility of the rats associating rewards with voluntary movements of muscles which might cause changes in the autonomic function. Under such highly controlled conditions, in which the vital processes of the animals were supported artificially, they used two methods of reinforcement to induce learning in the rats. One was to stimulate electrically a 'pleasure centre' of the hypothalamus as a reward, and the other was to cause avoidance by a mildly unpleasant electric shock. When the rats changed their rate of heartbeat slightly during the presentation of light and tone signals, they rewarded or punished them by these methods. One group of rats was rewarded every time the rate increased, while the other group was rewarded every time the rate decreased. Both groups of rats learned to alter the heartbeat as a result of the training (Fig. 90). Similarly, intestinal contractions, changes in blood

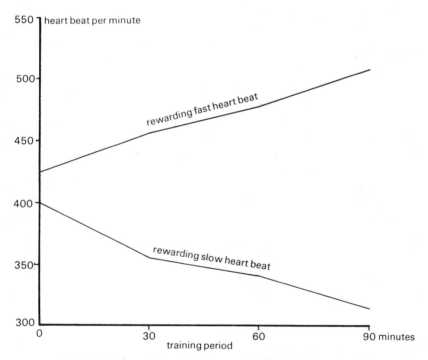

FIG. 90: Autonomic learning by curarized rats being rewarded with pleasurable brain stimulation. Rats rewarded for increasing the rate of heart beat and those rewarded for decreasing it showed a marked difference in the rate after a period of training. (AFTER DiCARA)

pressure and other visceral responses have been shown to be controllable voluntarily under certain conditions. Such learning ability indicates that classical conditioning and instrumental learning are not very different as used to be considered, and, if properly explored, the training may have a therapeutic value for cardiovascular or other visceral disorders.

TRIAL-AND-ERROR LEARNING

B. F. Skinner, an experimental psychologist at Harvard, developed a special apparatus which has become known as the *Skinner box*. It is a well-ventilated soundproof box containing an object for the animal to manipulate. Activities of animals in the Skinner box are automatically recorded. In a study of the learning ability of pigeons, Skinner and his associates fed the bird in such a box for a standard period on a diet carefully adjusted to produce a state of steady hunger. The object in the box was an illuminated disc which, when pecked, delivered a grain of corn. Pecking is an active response of birds to a variety of objects, and is therefore a useful character for learning experiments. After the long period of experimentation and frustration which is inevitably associated with the development of a new technique, the researchers produced a special food-delivering mechanism which can be adjusted to produce a grain of corn at every peck or only after 10, 100, 1000 or any desired number of pecks. Pecking activity was recorded automatically on a revolving drum, so that the total number and rate of responses could be obtained. Under such schedules of intermittent rewarding, two pigeons were found to respond persistently when only one in every 875 pecks was rewarded. When the conditioning was complete the experimenters stopped rewarding the birds and counted the number of pecks delivered. In the absence of reward, the pecking response appeared as many as 10,000 times before the disc was ignored. Moreover, it was found that some schedules of intermittent rewarding were more effective in producing pecking behaviour than was regular rewarding.

Under natural conditions rewarding is never regular, often being very infrequent and uncertain when animals hunt food. Thus, trial-and-error learning obviously plays an important role in the adaptive modification of their behaviour.

IMPRINTING

A famous German ornithologist, Oskar Heinroth, reared geese in isolation from the egg stage and found that young geese followed him wherever he went. This 'following' response appeared to establish social associations during the early part of an animal's life. Geese followed a man, or the first relatively large object they saw, just as they would follow their parents. They took the human keeper as both parent-companion and fellow member of the species and, at maturity, would even direct their sexual behaviour towards man. This type of early social learning is called *imprinting*.

Highly social species have a stronger imprintability than solitary species. Cuckoos reared in the nests of foster parents are generally indiscriminate in their food-begging behaviour and, as soon as they become independent, stop following the foster parents. If hand-reared, they show no social behaviour other than food-begging and leave the keeper when they fledge, just as they would their natural foster parents. On the other hand, Australian finches are very social birds and, if hand-reared in isolation, can be made to associate with man and not respond to behaviour and calls of their own species. A hand-reared male zebra finch sings courtship songs at the sight of a man and attempts to sexually mount his fingertips and ear lobes. If young birds are reared together, however, they will imprint on each other, so that innate social behaviour will be directed towards their nest-mates. The period in which an individual may become imprinted seems to vary from species to species, and often is affected by the maturing of instincts, as well as by other forms of learning such as habituation and conditioning. The development of fear response, for example, inhibits the ability to imprint, so that under natural conditions young become imprinted on one kind of moving object—their parents. If fear responses are reduced by drugs or habituation, the sensitive period for imprinting may be extended. All innate social behaviour requires recognition of members of the same species and since imprinting seems to contribute much to this function, it is of great significance to the study of social maladjustments and misorientations. At the same time, the manifestation of social instincts depends on this form of learning, especially in many higher animals with well-developed sociality and parental care. Thus the study of imprinting has a promising future in revealing the releasing mechanisms of innate social behaviour.

Harry Harlow, who leads a team of psychologists at the Primate Laboratory of the University of Wisconsin, studied the infant-mother reaction in the laboratory by depriving newborn rhesus monkeys of their real mothers and providing them with substitute mothers made of various materials. When offered a choice these monkeys showed an innate preference for a form covered with soft terrytowelling over one consisting of a bare wire-frame, even when a milk bottle was provided on the wire frame only (Fig. 91). In their adult life they showed abnormal responses to the opposite sex, not only among themselves but also to normal individuals which were known to have mated successfully in the past. Many of the motherless female monkeys did not succeed in bearing progeny and those that did proved to be very poor mothers, ignoring or ill-treating their own young.

167

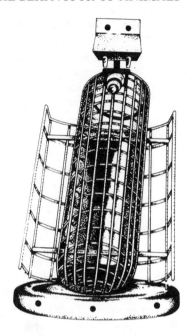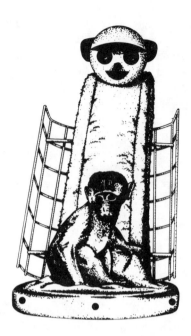

FIG. 91: An infant rhesus monkey with two artificial mothers. The amount of time the monkey spent on each was recorded automatically by its weight activating a mechanism in the pivoted 'mother'. The monkey chose for comfort the one with a covering of soft towel (right) in preference to the one with a wire frame (left), though the latter supported a milk bottle. When frightening objects were introduced into the cage, it would run to the cloth mother to press and rub against her. Then it would look at the strange object while still keeping physical contact with her, as shown in the diagram. (DRAWN FROM HARLOW)

These results show that lack of imprinting in early life affected reproductive behaviour in later life. Among wild monkeys, dependent young will not survive if deprived of their parents as they are rarely adopted. Natural selection, therefore, must operate strongly against the inheritance of any slight abnormality in reproduction.

LATENT LEARNING

During flights to and from the nest, the female sand wasp *Ammophila* learns important features of its environment. This learning is not in relation to the particular motivation which causes the flight and therefore it has no immediate use for it. However, when the wasp transports caterpillars that are too large to carry in flight, it drags the prey back to its nest using the 'knowledge' of surroundings that it gains during the flights. The wasp

168

can walk back to the nest even when forced to detour, or after displacement of up to 40 m.

This type of learning, called *latent learning*, occurs without reward at the time of learning, and the stimuli or situations learnt remain irrelevant until later put to use. Thorpe offered the larvae of *Drosophila* flies food soaked in peppermint oil. The adult flies which had received this treatment during the larval stages preferred the odour of peppermint while searching for food. This is a conditioning in which an association has been established between food and a particular odour, but because food always contained peppermint oil the association was irrelevant during the larval stages. In their adult life, however, the flies under hunger motivation used this 'knowledge' in trying to locate food. This process resembles imprinting in vertebrates.

Homing in a number of molluscs often involves a 'knowledge' of the topography around the homesite gained during feeding trips out from the homesite. The large Barrier Reef chiton *Acanthozostera gemmata* occupies a well-marked homesite on beach rock during the day (Plate IIIa). At night it emerges to feed, browsing a trail through the encrusting algae. To return to its homesite the chiton follows its out-browsed trail. Although under normal circumstances this is the way the chiton homes, it can also find its way back to the homesite when experimentally displaced. But it can only do this if it is displaced within its normal browsing range. There are a number of mechanisms which could be employed in homing after displacement; for example, the chiton might detect the odour of its homesite, or follow 'old' trails back. But a number of experiments have indicated that chitons may have a topographical memory and therefore can utilize this information when displaced.

If a mollusc does not follow its out-browsed trail or does not home using some beacon such as sight or smell of the homesite, this is usually taken as evidence of latent learning; i.e. the animal has learned features of its environment. The limpets *Acmaea* and *Patella* can home without using their out-browsed trails. R. Ohgushi in Japan found that the lunged limpet *Siphonaria japonica* in a new area always homed along its outgoing trail. However, with increasing time spent in a particular area the limpet became less dependent on this method of homing; it had apparently become 'familiar' with its surroundings.

INSIGHT

A captive monkey, on seeing attractive food outside his cage, will try to reach it from inside. He may first extend his arm, trying to grab the food from several directions (this is trial-and-error). If unsuccessful, he may find a stick or some other long object that he can hold and use to obtain the food. In this new behaviour, an object is manipulated as though it were an extension of his arm. In other words he uses it as a tool, 'knowing' that it will reach the food (showing purposiveness in behaviour); trial-and-error experiments may have been conducted mentally without any visible actions. This ability to organize behaviour from past experiences is termed *insight*. An experienced dog can make a detour round an obstacle without preliminary random trial-and-error behaviour. This *detour behaviour* suggests the presence of insight. Another response, *imitation*, also suggests insight. In imitation, the act is novel to the animal imitating it and does not arise from instinctive tendency.

In human learning the solution of complex problems comes from *reasoning* and *symbolic behaviour*. Rudimentary forms of these are found in higher vertebrates, which show insight rather than trial-and-error in the solution of problems.

When an object or situation is recognized and retained as a generalized image in the memory, we say that *perception* has occurred. Anticipation of encountering the object or the situation again will help confirm or modify the image when the particular encounter recurs. This process is directed socially, particularly through education. Education involves encouragement of interest and self-directed activity in children, as learning is limited and inefficient when there is no interest in the subject.

Perception involving visual form and pattern plays an important role in developing adaptive behaviour in babies. Robert Fantz, an American clinical psychologist, measured the responses of two to three-months-old babies to various round figures shown in Fig. 92. He found that visual perception is guided to some extent by the innate tendency for association with particular forms. Colour or brightness is not as important as pattern.

If you try to describe houses along one section of a street between your house and town, you may not be able to recall them all, although you have seen these houses a number of times in the past. *Memory*, therefore, is not merely retention of the impression gained through experience, but involves its recall into awareness.

Some information received may be stored without ever being brought into consciousness, but such information is generally lost as forgetting takes the upper hand. Information acquired through more conscious effort can also be lost. Otherwise examinations would not worry students.

We tend to retain memories of vivid impressions, repeated experiences, and experiences undergone in youth. It is now generally accepted that short-term memory and long-term memory have different nervous mechanisms. Lloyd Peterson, an American psychologist, was curious about the workings of memory and so carried out many experiments with people.

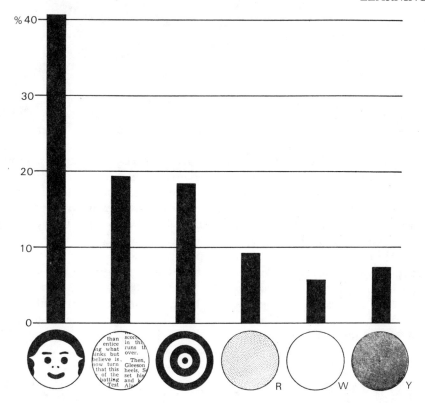

FIG. 92: Results of experiments to assess the relative importance of pattern, colour and brightness in the response of two to three-months-old babies. The percentage of the total gaze by infants was greatest for a face, followed by a piece of printed matter and a bullseye, showing the importance of pattern over colour and brightness, represented by plain red, white and yellow discs. (AFTER FANTZ)

One of these was to introduce words in pairs, one pair at a time. Every now and then the subject was asked to recall the word associated with one presented earlier. Using paired words similar to those shown in Fig. 93, he obtained interesting results. When words were shown at a rate of one pair per two seconds the recall after one intervening exposure was better than the recall following exposure at the rate of one pair per four seconds. But the recall after two to six intervening exposures was better at the four-second rate than at the two-second rate.

If you run a similar test yourself, you will find that learning occurs faster and memory lasts longer if some association of meanings exists between the paired words. Similarly, you may be able to examine the reasons why you can recall some terms and not others in this book.

tooth—book   fence—[＿＿＿]   tree—face   egg—[＿＿＿]

fence—apple   arm—moon   arm—[＿＿＿]   hair—door

car—pear   table—eye   egg—worm   table—[＿＿＿]

FIG. 93: Peterson used paired words to test recalling abilities. Subjects were asked to recall the associate of a particular word (e.g. 'fence', 'arm', 'egg', 'table' in the illustrated pairs) immediately after exposure (e.g. 'egg'—'worm') or after one to six other associated pairs.

Man can reason far beyond the capacity of other primates. Formation of concepts and principles, however, is evident in non-human primates and in some birds (e.g. crows). Apart from detour and other similar behaviour suggesting the presence of insight, crows can improve their learning capacity by performing a large number of similar tasks. Also, they can learn to choose an odd object out of three after training with several sets of entirely different objects (Fig. 94). This kind of learning requires transfer of training, involving formation of principles. On the other hand, counting and learning of a particular sequence involves the capacity to symbolize objects. Both reasoning and symbolic thinking are important in the formation of scientific principles. From these capacities man originally developed tools and languages.

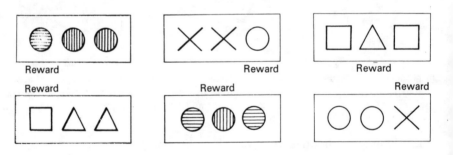

FIG. 94: Oddity test: some monkeys and crows can learn to select the odd pattern out of three, two of which are the same. They learn that the odd one is rewarded regardless of where it may appear and that it is the odd pattern, not necessarily the same pattern as one rewarded previously, that will be rewarded, as shown in the diagram. This type of learning involves formation of principles.

8 Social Organization

We have seen that there are many types of innate and learned behaviour in animals. Let us now examine how social behaviour is organized.

Animals responding only to physical factors of the environment may be found in a group. For example, soil arthropods or protozoans may collect together in suitable conditions of humidity, pH and temperature. Such groups are called *aggregations*. If, on the other hand, members of a group stay together as a result of their responses to fellow members, the group is regarded as *social* and such responses are called *social responses*.

SOCIAL BEHAVIOUR

Escape from predators or parasites is important for survival, while exploitation of food and other resources is important for both survival and reproduction. For successful reproduction, selection of a suitable mate and protection and care of the young are also important, particularly in higher animals. In the course of these life processes, interactions occur between individuals of a social group. It is thus understandable that both competition and co-operation developed in the social behaviour of animals.

COMPETITION

Competition does not necessarily involve fighting or even contacts between individuals. Plants competing for nutrients and water in the soil and for sunlight do not normally have contact. Similarly, larval oysters settling on a rock surface compete for available space without fighting.

The requirements of life often overlap between different species of animals and, as a rule, the closer their evolutionary relations the more similar are their requirements. However, among closely related species, particularly if they occur together in the same area, there are often subtle but significant differences in their requirements or in their methods of exploiting limited resources.

Competition may occur between species or between individuals of the same species. In most animals which produce more young than necessary to replace their generation, the population increase is checked at some stage by various factors of the environment. One of these factors is social interaction.

Young birds compete for food that parents bring to the nest. If food is in short supply, some nestlings die of starvation. Parasitic cuckoos, when they hatch, instinctively push the eggs of the foster parents out of the nest. If the nestlings of the host are already present when the cuckoo hatches, it will still try to throw the nestlings out of the nest. It will also monopolize the food brought back by the foster parents by using a strong begging response.

Honeyeaters are aggressive birds, often seen chasing one another while feeding in a flowering tree. Although all local species of honeyeaters may aggregate on the same flowering tree, there is a definite order of social dominance among the species. The abundance of each species, however, is related to the abundance of particular species of flowering trees, so that the habitat is often divided among the different species. Flying foxes feed at night often from the same flowers visited by lorikeets and honeyeaters in the daytime. Thus direct interference is avoided, although competition may still occur between the day feeders and night feeders.

The form of fighting between individuals of the same species varies from mild threat displays to severe fighting, depending on the motivations of the animals involved and on the circumstances causing fighting. The behaviour used in fighting consists of postures and movements signifying the outcome from the complete victory to the complete defeat. Such social behaviour, including both attack and escape responses, is called *agonistic behaviour*. The meaning of 'agonistic' covers the behaviour of both winner and loser in a contest.

CO-OPERATION

In spite of the increased interaction among individuals living in a group, many animals form social groups in which the activities of individuals are integrated or regulated. This occurs because social living affords many advantages to the members of the group. Co-operative association of individuals may occur either between species or within species.

Remora fish (sucker fish) are transported to feeding areas by sharks without benefiting or harming the sharks (*commensalism*). Sponges and crabs living in the cavities of sponges benefit from each other without obligatory relations (either can live without the other). This is an example of *protocooperation*. In the case of flagellate protozoans living in the intestine of termites, each depends on the other for its existence (*mutualism*). These examples of co-operation between different species are called *symbiotic relations*.

An interesting relation is found between algae and corals. For a long time it was considered that the algae zooxanthellae did not provide nutriment

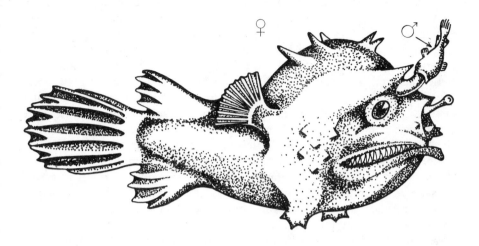

♀   ♂

FIG. 95: Sexual parasitism in the deep-sea angler fish *Photocorynus spiniceps* (AFTER NORMAN): the male is nourished by the blood of the female to which he is attached

for the corals. These algae can resist digestion and grow in the tissues of giant clams and corals. If the coral is starved or kept in continuous darkness, the algae will be extruded rather than digested. However, subsequent studies of the energy relations of a coral reef indicated that corals would have to be obtaining some food from the algae if the energy budget of the reef was to be balanced. Further experiments demonstrated that zooxanthellae can live outside the coral as normal Dinophyceae (dinoflagellates) and that they produce a soluble organic substance, glycerol, only in the presence of living coral tissue. Thus it now appears that from the coral the algae receive nitrogen, phosphorus and carbon dioxide required for photosynthesis, as well as protection; the coral in turn receives glycerol for nutrition from the algae.

Co-operation between individuals of the same species varies again from the non-obligatory association, such as herds of grazing animals, to the obligatory association, such as the colonial hydrozoans. These associations generally serve one or more of the following functions in the life of animals.

● REPRODUCTIVE FUNCTIONS: An extreme case is *sexual parasitism* in which the difficulty of one sex finding the other is met by permanent attachment of the small male to the female, as in the deep-sea angler fish (Fig. 95). Only the female catches food, and the male is nourished by her blood.

175

Sometimes, external factors bring many scattered individuals together for breeding. For example, lunar cycles govern the breeding activity of some marine polychaete worms which spawn in large numbers during the spring tide.

In higher animals, social responses bring the sexes together. They may also depend on external stimuli for timing, as in the case of trout and frogs which congregate in the breeding grounds. Mating flocks of swifts and breeding colonies of water birds, sea birds and seals are further examples of social groups having reproductive functions. A rather unusual example of co-operation is seen in the bent-wing bats, *Miniopterus schreibersii* (Plate IIIb), which form nursing colonies in special caves. In the temperate region of eastern Australia, the nursery is usually established in a blind chamber of a humid cave, connected to the outside only through a narrow passage. In such a cave the heat insulation is nearly perfect. Peter Dwyer observed about 10,000 females in one of these caves. They arrive several weeks before giving birth to their young, and warm the chamber by active flights and through dissipation of general metabolic heat. The temperature of the cave is normally about 15°C, but as a result of this activity it rises to about 27°C, which is suitable for nursing the naked young.

● SAFETY FUNCTIONS: These are concerned mostly with escape from predators. The escape response of animals in a group is either to scatter, as in the case of a school of small fish being chased by a large fish, or to tighten the group, as in the case of a flock of starlings meeting a hawk (Fig. 96).

Social animals have alarm systems to warn the members of danger. Thumping by rabbits and kangaroos and shaking of a tree by monkeys are often understood as warning signals. Many birds have alarm calls which are recognized by fellow members of the group and sometimes also by other species in a mixed flock.

● CONSERVATION FUNCTIONS: These are concerned with utilization of food resources. In extreme cases, division of labour is established, so that individuals perform specific tasks for the benefit of the group. Ants and termites developed morphological differences for the specific tasks of individuals involved (*castes*). An interesting example of sexual co-operation for food hunting was provided by the huia (Fig. 97), a New Zealand bird now believed to be extinct. The female had a long and slender bill, curved sharply towards the tip. The male had a stout bill, about two-thirds of the length of the female's bill. They hunted together for insects in decaying logs and green saplings. The male would first tear away the outside and chisel out much of the wood; the female would then probe inside with her slender bill to obtain the prey.

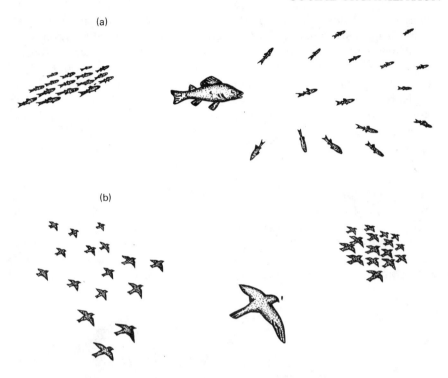

FIG. 96: Escape responses of social groups: (a) small fish scatter when a predator appears suddenly, and (b) starlings in flight tighten their formation when a hawk attacks

Social animals have a better chance of finding good food sources than solitary animals, because the effort is made by a group of individuals. Competition for food, however, will offset this advantage of social living. Thus, any reduction in competition will be favoured in the evolution of social groups. This seems to have been achieved in several ways. For example, the group size may be kept small, so that the demand for food will never be too great; when the group becomes too large, it will split into smaller groups.

The general health of individuals depends on their energy balance. Therefore, fighting for food is not advantageous, particularly when the amount of food that becomes available as a result of fighting is not sufficient to compensate for the loss of energy involved. Also, if little or no energy is lost through fighting, then food can be better utilized by the whole group. Thus, normally territorial animals may cease to be territorial if they become overcrowded, and loss of energy through frequent fighting will be avoided.

Similarly, avoidance behaviour of submissive animals will be advantageous when food is limited in distribution but not in quantity. They can

177

obtain food without fighting if they wait until dominant animals finish feeding. Aggressive animals may be better able to survive than submissive animals when there is a shortage of food. However, if severe fighting should develop over a very limited amount of food, aggressive animals might spend more energy than they could obtain from the food. Thus, energy conservation is best achieved if *dominant-subordinate* relations are established between individuals, who can recognize one another in the group and avoid unnecessary fighting. In such a group dominant animals are not always aggressive. For this organization to be effective, the group size must not be too large. If the group size is too small, however, the group will lose its safety functions.

It is interesting that the size of feeding flocks of birds is generally small when birds utilize food which is scattered or limited in distribution. On the other hand, in non-nutritional activity such as migration, the flock size may become enormous. If each species utilizes a different food resource, the safety functions of mixed flocks can be increased without intensifying competition. Members of such flocks will respond to warnings produced

Fig. 97: The huia *Heteralocha acutirostris*, an extinct New Zealand species, showing sexual dimorphism in size and shape of bill (REDRAWN FROM OLIVER)

by others at the sight of a common predator. In such flocks conservation functions will remain effective, or sometimes even increase if the activity of one species makes resources for other species more readily exploitable. For example, fruit-eating birds often disturb foliage insects which become easier prey for flycatchers. Such mixed flocks of birds actually occur in tropical rainforests of south-east Asia and Panama.

Wolf packs show a high degree of organization and co-operation when systematically hunting other animals. Beavers build dams, living quarters and canals to ensure a constant supply of water and to facilitate reproduction and transportation of food supplies. In troops of monkeys, division of labour is established according to the sex and the age group, but the task of each group may be different in different situations. Thus, social groups of mammals often serve many functions and individual members may be organized to perform different tasks.

COMMUNICATION

Social responses of animals are based on communication between individuals. We shall now examine the messages contained in communication.

Language in our society is not the only means of communication, as various signals used between the communicator and the recipient may not form words. Signals are contained in the expressions and postures of the communicator, who produces messages through behaviour either consciously or subconsciously. This process is enormously magnified in animal communication.

LEVELS OF COMMUNICATION

All social responses are, therefore, the result of communication and all social behaviour involves communication. The communicator and the recipient are necessary for communication to take place, but the communicator is not necessarily aware of the signals he is sending.

Animals must be able to recognize their own species, and in *species recognition* the signals are not directed to any particular individual. Such signals persist for as long as the animal is being seen, or is producing sounds or other signals of the species. If behaviour is involved it is innate, though the recognition may be based on imprinting. Recognition at the species level must be particularly efficient in those animals that are sparsely distributed, those that mimic other species, and those that associate with similar species. Different species of Australian finches often flock together, but each species or race has distinct markings on the body (Fig. 98 and

179

insignificant uniform colour ☐
white ●
black ○
barred ▤
spotted ⠙
red ▦
yellow ▲
brown ⧖
blue ■
green ▨
grey ▭
red with black line ⊟

FIG. 98: Body areas having distinct markings of races and species in Australian finches (Estrildidae): (1) beak, (2) lores, (3) eyebrow, (4) ear patch, (5) crown, (6) back and wings, (7) rump, (8) tail, (9) throat, (10) breast, (11) belly, (12) flanks, (13) legs. On the next page, note the degree of similarity in colouration between races of the same species (marked by the dotted line on the right) and between species of the same genus (marked by the solid line on the right).

Plate IVa and b). Curiously, no species has special markings on the wings or back, but the colour of the rump feathers may be used for recognition of the species during flight.

Social animals generally recognize members of their group by sight, odour, vocalization and/or particular behaviour. This is *group recognition*. This level of communication concerns cohesion of the social group and produces antagonistic reactions to strangers. Social insects (e.g. ants) distinguish strangers by odour, and attack them. In higher vertebrates, group members are often identified by *individual recognition*. Social groups often consist of various age and sex classes, and communication among classes organizes the group into a particular society. Stuart Altmann measured the amount of interaction within each class and among various classes of rhesus monkeys. He found that the pattern of interaction did not fit a mathematical model dealing with random frequencies of communication. Since little interaction means little communication, the classes can

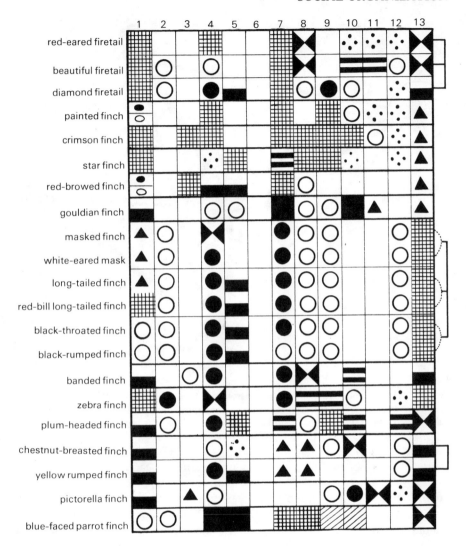

be arranged from the highly communicative central class to the least communicative peripheral class, as shown in Fig. 99. Non-random interactions of this kind suggest that messages exchanged are different according to which classes are involved in communication.

Sexual recognition is necessary for the reproduction of most bisexual animals, but this may or may not be based on individual recognition. If sexes are similar in appearance, behaviour (sound, scent or movement) indicates the sex, at least during the reproductive phase.

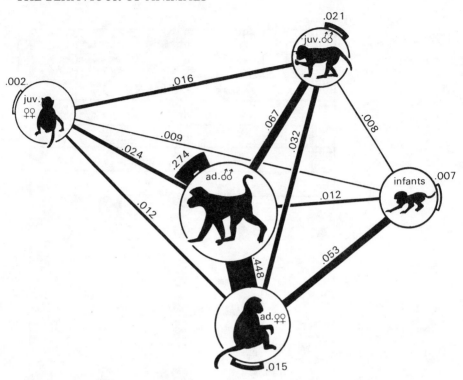

FIG. 99: A model of social interactions between and within classes of rhesus monkeys. Probabilities of various interactions, shown by the figures and the thickness of the lines, are calculated from a hypothetical population with equal representation of age-sex classes. (AFTER ALTMANN)

In studying communication it is important to learn the circumstances under which a particular communication takes place, the identity of communicator and recipient, the message and meaning transmitted, and also the functions served by each of these. The message conveys what the communicator 'intends' to do, and the meaning informs the recipient of what he should do. For example, the message contained in the threat posture of an animal may be 'I am going to fight back', whereas its meaning may be 'Don't attack'.

METHODS OF COMMUNICATION

The means of transmitting messages often differ according to the species, caste or class involved and the context in which the communication occurs. They may involve sight, sound, odour, taste, touch or patterns of behaviour. In various displays some of these are combined to produce maximum effects.

Many animals are equipped to produce sound vocally or by rubbing together hard parts of their body. When males produce sound to attract females, the sounds are often musically complex and are referred to as *songs*. Although the song may transmit other messages, such as territorial signals, of all the sounds the animal produces it is most distinctly characteristic of the species and often has qualities unique to the taxonomic group to which it belongs. For example, the frequency of sounds produced by cicadas forms a comparatively low and narrow band; clicking sounds produced by crabs have a low frequency; croaking of frogs has repeated pulses of wide frequencies, and bird songs consist of melodious, harmonic sounds. These are illustrated in the sonograms shown in Fig. 100.

Interesting methods of communication are found in the honey bee, which has been studied in detail by von Frisch. Bees returning from a new food source dance in the hive and the dance is repeated by other workers of the hive. During the dance, information is given about (a) the type of food available, (b) its abundance, (c) how far away it is, and (d) its direction with respect to the hive.

(a) The odour of the site visited, usually flowers, adheres to the body hairs and is detected by fellow workers as they follow the dancing bee and touch it with their antennae. They receive nectar samples from the forager and in this way obtain information relating to the type of food source and the kind of flower to visit.

(b) Dances normally range from a few seconds to about three minutes, but if a particularly rich source of food has been located the dancing bee will increase the *duration* of its dance and so recruit more workers. In addition, the dance is performed with more vigour or 'vividness'.

(c) The distance to the food source can be correlated with either the type of dance displayed or the way in which it is performed. For the Austrian bee *Apis mellifera carnica*, a forager locating a source closer than about 85m from the hive performs a *round dance* (Fig. 101a) on returning to the hive. Apparently the distance information of this dance is that 'Somewhere within an 85m radius of the hive, food is available'.

If the food source is greater than 85m from the hive, the Austrian bee forager performs a *tail-wagging dance* (Fig. 101b) on returning to the hive. A number of features of the tail-wagging dance can be correlated with distance; perhaps the best correlations are (i) the duration of a single tail-wagging phase, or (ii) the number of tail wags during the straight run. But other features may also be correlated with distance, for example the speed or length of the wagging run, or the duration of the semicircular run.

Which of these features, *if any*, are used by the recruited bees is not clear. What is clear is that humans can observe the dance and can relate these features to distance. To prove the accuracy of bee communication, von

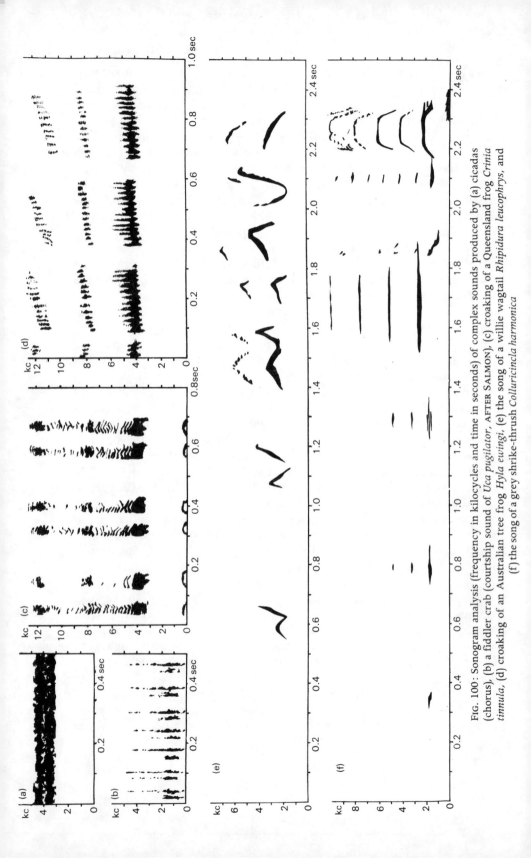

Fig. 100: Sonogram analysis (frequency in kilocycles and time in seconds) of complex sounds produced by (a) cicadas (chorus), (b) a fiddler crab (courtship sound of *Uca pugilator*, AFTER SALMON), (c) croaking of a Queensland frog *Crinia tinnula*, (d) croaking of an Australian tree frog *Hyla ewingi*, (e) the song of a willie wagtail *Rhipidura leucophrys*, and (f) the song of a grey shrike-thrush *Colluricincla harmonica*

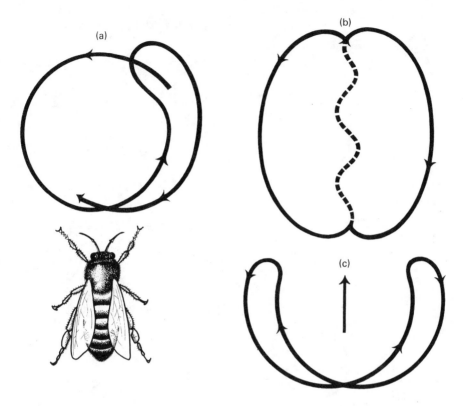

FIG. 101: Dancing patterns of honey bees: (a) round dance, (b) tail-wagging dance, and (c) sickle dance. The straight run with abdominal wagging is indicated by the broken line. In (c), the arrow indicates the direction of the food source. (AFTER VON FRISCH)

Frisch asked his daughters to build a feeding table at a site unknown to him. They chose a spot 342m from the hive over hills and woods. After observing the dances of individually marked bees returning from the table, von Frisch was able to estimate the direction and distance to the table within a few metres of the actual position.

Other races of *Apis mellifera* (but not the Austrian race) show a third type of dance, the *sickle dance* (Fig. 101c). This is used to indicate distances intermediate between those of the round and tail-wagging dances; for example, the Italian bee *A. mellifera ligustica* dances the sickle dance when the distance to the food source is between 10m and 40m. There is, in addition to the distance information, an indication of direction in the sickle dance: the food source is on a line bisecting the opening of the sickle shape.

(d) An indication of direction is given by the direction of the tail-wagging run. The angle of this direction with respect to the anti-gravitational

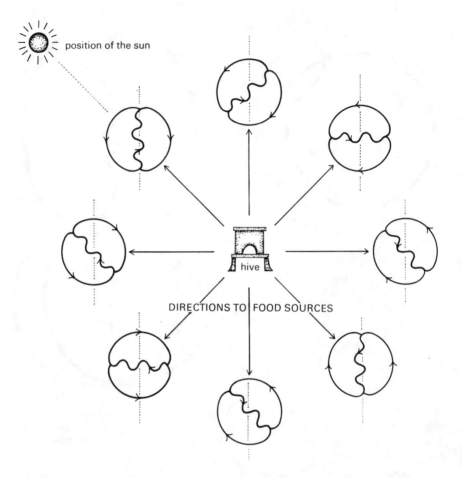

FIG. 102: The orientation of the tail-wagging dance of the honey bee on a vertical comb in relation to the direction of food sources

direction is equal to the angle formed between the sun's position and the food source (Fig. 102).

Harald Esch spent some time studying bees with von Frisch. He had been puzzled that in the total darkness of the hive the bees could presumably observe the dance and interpret its tempo as indicating the distance to the food source. He built an artificial bee which was driven by a motor to perform the tail-wagging dance in the hive. The bees in the hive clustered around the dummy just as they would around the genuine forager. No bees, however, left the hive to search for food.

186

He thought this might be because the dummy did not vibrate in the same way as the live bee during the dance, and therefore he recorded sounds produced by the dancing bees. The microphone revealed that during the tail-wagging dance the bee produced strong sound signals with vibrations of its wings. The period of sound production coincided exactly with the duration of the straight run of the dance. Since von Frisch had found the correlation of the latter with the distance to the food source, Esch concluded that bees used sound as the cue to distance.

Esch also described how some of the advanced species of stingless bees, *Melipona*, use sounds to indicate distances. Short-duration sounds indicate a source close to the hive, while longer drawn-out sounds indicate sources at 100m or more. Direction is indicated by the forager leading a group of alerted bees along a zig-zag path in the general direction of the source. Then, after about 20-30m, the leader flies direct to the food source while the followers go back to the hive and wait at the entrance for the leader's return. More food-sharing then occurs and the group flies off again, only to return once more, leaving the leader to visit the source. This procedure continues for a while until eventually some of the group fly to the food source on their own.

Other species of stingless bees, e.g. *Trigona postica*, recruit workers by running about noisily and giving out food samples. When a number of workers have been alerted, the forager leads these bees along the scent trail. Scent marks are deposited every few metres on stones or vegetation on the way back from the food source.

The dwarf honey bee *Apis florea* dances only on a horizontal surface and appears incapable of transposing a horizontal dance to a vertical one. The ordinary honey bee sometimes dances at the hive entrance on sunny days. In this horizontal position they indicate direction in the same way as the dwarf bees; the tail-wagging run of the dance points directly to the food source.

The methods of communication that honey bees use are the most complex of all the various types of communication known in the invertebrates. However, we can observe behaviour patterns in other groups of insects which show similar aspects to those of the honey bee. They may indicate possible evolutionary paths that bee communication has taken.

A number of non-social insects perform stereotyped movements after feeding or flying, but the information content of such movements is usually 'wasted' as it is not communicated. For example, rhythmic swaying movements are shown after flight by the saturnid moths, and the number of such movements depends on the distance travelled. Coccinellid larvae make many small turns immediately after feeding on an aphid (see p. 131). In this case the movement serves the purpose of keeping the animal near the possible food source but other individuals do not make use of it as a signal.

V. G. Dethier has demonstrated that the behaviour of the fly *Phormia regina* shows many similarities with that of bees. After feeding on a source of food which is insufficient to satisfy, flies perform repeated turnings in either direction. The rate of turning and the duration of such movements depend on the richness of the food source, in much the same way that the vividness and duration of bee-dancing depend on the concentration of the nectar located.

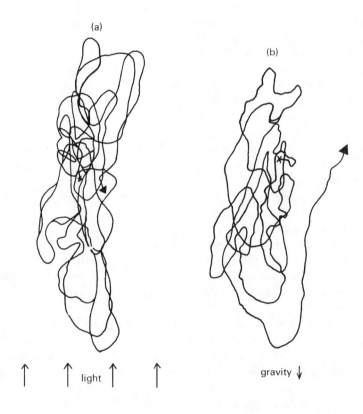

FIG. 103: Brief stimulation with sucrose causes a fly *Phormia regina* to produce (a) an elongated pattern of locomotion on a horizontal surface in a beam of light. A similar pattern (b) is shown if the fly crawls on a vertical surface in darkness, as though it were transferring the direction of light to the direction of gravity.
(AFTER DETHIER)

Unfed flies become excited when a recently fed fly regurgitates some of its meal. They follow the fed fly around, and begin to make turning movements. The food-sharing activities of honey bees, where most of the collected food is regurgitated and transferred to other workers, could have evolved from such a behaviour pattern. If a stimulus such as directional light is present, the 'dance' of flies acquires a directional component and the turnings are elongated parallel to the light rays (Fig. 103). In darkness, the dance on a horizontal surface lacks a directional component, but this factor is present when dancing on a vertical surface, the turnings being elongated parallel to gravity. There are similarities here with bees, whose dance is oriented with respect to light on a horizontal surface and with respect to gravity on a vertical surface in darkness.

Dung beetles walking on a horizontal surface in a darkroom will, if unilaterally illuminated, maintain a straight path at a constant angle to the source. If the surface is tilted to vertical and the illumination switched off, the path travelled remains straight and at an angle to gravity equal to that shown earlier towards the light. Bees have this sort of ability; the tail-wagging portion of the dance in the hive is at the same angle to gravity as was the sun's position with respect to the flight-path back to the hive.

The whole concept of bee communication, accepted for many years, has recently been questioned. Adrian M. Wenner and Dennis L. Johnson point out that bees which have never before visited a particular food source do not find it easily; only relatively few of those bees recruited from the hive are successful in locating the food source. Besides, the successful ones take a long time to reach the source, even up to twenty times the time needed to fly a straight line between the hive and the source. They suggest that these inexperienced foragers locate the source by dropping downwind from the hive and then searching for the right combination of odours from food, locality and other bees.

Bees feeding at a particular food source remember aspects of the situation. If later a successful forager returns to the hive with a particular food and/or locality odour, the workers can utilize their 'knowledge' of the hive environs and are recruited to the source. Thus Wenner and Johnson claim that bees are recruited by means of a conditioned response to food or odour, rather than by using information on distance and direction coded in a dance. They suggest that the 'experienced' foragers provide the major food supplies for the hive.

Since the experiments of von Frisch have thus been challenged, the solution to this problem must await future experimentation. If the interpretation of Wenner and Johnson is correct, we are still posed with the question of why bees provide information in their dances which can be interpreted so precisely by man.

CASTE SYSTEM

Social responses of individuals are organized into what we recognize as organizations or structures within the group.

In a highly developed society of insects, such as ants, termites, some wasps and bees, there are several castes (Fig. 104) which differ in both structure (showing *polymorphism*) and function (showing *division of labour*).

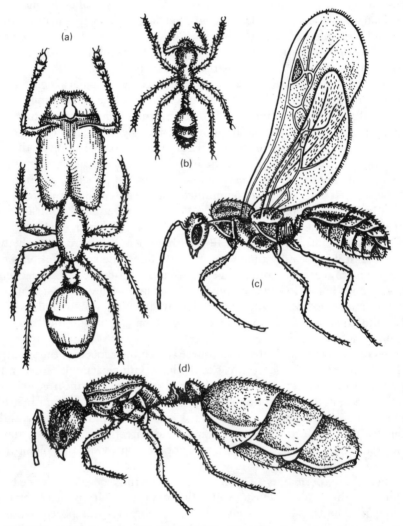

FIG. 104: Different castes of ants *Pheidole instabilis*: (a) soldier, (b) worker, (c) reproductive caste (male), and (d) reproductive caste (wingless female) (AFTER WHEELER)

The reproductive caste is concerned only with colony establishment and production of other castes. The worker caste is generally sterile and concerned with the collecting of food, feeding of others, and maintaining the nest. The soldier caste is often equipped with large jaws and guards the nest and protects other castes.

In most social insects the reproductive males do not perform any social activities other than fertilization, whereas the females often build nests, lay eggs and care for the brood. In termites and some ants, wingless males labour as workers, but in other social insects it is the females that differentiate into various castes.

The question of how the different castes are produced from the genetically uniform larvae, in such proportions as are necessary for the effective organization of the colony, is not clearly understood. The distribution of juvenile hormones is probably related to the differential caste production, whereas the inhibition of ovary growth in workers is effected by a pheromone produced by the queen (see p. 64).

In spite of the complex methods of communication developed for conservation functions, the honey bees have no co-ordinating centre or messengers in the caste system. The division of labour is dependent upon physiological factors associated with the bees' age. When workers mature they first work as cleaners of cells. In their second week of adult life they produce wax and become the builders of cells. The pharyngeal glands become active at about the same time, and they nurse young by receiving nectar from foragers and putting it into cells. As their secretion diminishes, they spend the third week as guards at the hive entrance before becoming foragers for the rest of their lives.

In the case of emergency, however, they change diet and become physiologically adapted to whatever task is required for the continued existence of the colony. They patrol the hive frequently and discover any situation demanding attention. Their response to the emergency situation is thus entirely individual and is not organized by social co-ordination.

In higher animals, different sexes and age groups, if they assume different functions, may be referred to as castes or classes. These do not form such rigid caste systems as those of insects.

TERRITORIALITY

As seen in sticklebacks (p. 152), the social response of defence against other individuals, particularly of the same species and sex, leads to territorial behaviour if the defence is restricted within a certain area. As a result of territorial behaviour, individuals are often spaced out. This spatial arrangement is called a *territorial system*.

Animals may be attached to a particular site for varying lengths of time to satisfy specific requirements of life. It may be a homesite, a feeding spot or a display ground. They tend to be more efficient in gathering food and breeding in an area they know.

Greg Gordon studied movements of the short-nosed bandicoot *Isoodon macrourus* for a project at the University of Queensland. He placed a small radio-transmitter on selected animals and tracked them throughout the night when they were active. An example of movements detected by radio-tracking is shown in Fig. 105. The animal's activity was restricted within a definable area. Such an area, traversed by an animal in its normal

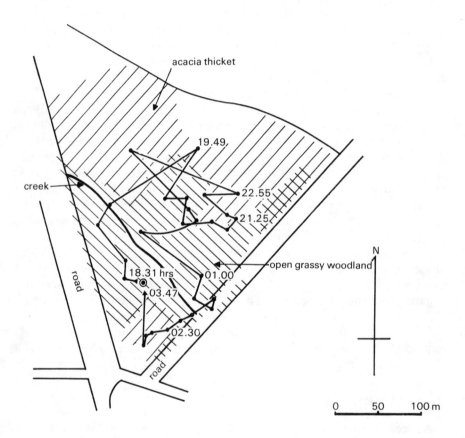

FIG. 105: Gordon studied the home range of the short-nosed bandicoot *Isoodon macrourus*. An example of home range activity shown here was revealed by overnight tracking of a female carrying a small radio-transmitter.

activities, is called a *home range*. If part of it is defended, the defended part becomes a *territory*. Thus, territoriality is a result of two tendencies: *site attachment* and *hostility*. Home range implies only the site attachment whereas social hierarchies may result only from agonistic behaviour without reference to the site.

John Stimson recently discovered that the owl limpet *Lottia gigantea* occupied territories on the rocky shore near Santa Barbara. Larger owl limpets had larger home ranges, but on the average the limpet occupied a patch of algal film about 900cm^2 in area. Within home range the limpet responded to three categories of intruders: competitors for food, competitors for space, and predators. Not all of them were kept out of the territory at all times, but occurrences of these animals within the territory were very rare compared with the area outside it. An introduced *Lottia* usually turned and moved away after contact with the resident *Lottia* who followed in pursuit. If the intruder was smaller than the resident it was caught up quickly and dislodged; if the intruder was about the same size as the resident it usually escaped. The dislodging was effected by shoving with a sudden thrust of the shell and foot at the intruder. The same method was used against other species of smaller limpets (e.g. *Acmaea*)—competitors for food. Owl limpets also prevented settlement of mussels and sea anemones in their territories by shoving them in the same manner, but did not respond to foreign objects placed within territories. They sometimes used a different method to drive predatory snails away; instead of lowering the shell the limpet raised it and brought it down quickly on to the anterior edge of the foot of the snail which then would retract its foot, lose its grip, and be washed away. Stimson's report is the first description of territorial behaviour in limpets.

Insects and spiders often show territorial behaviour, but in most cases the stronger individual wins the territorial fight. In higher animals, territorial fighting is rare. Instead, they exhibit threat postures and challenge calls. There is also an effect of prior residence in higher animals, and territorial encounters are normally won by the territory holder regardless of the intruder's strength. The size of the territory varies according to the function of territorial defence. It may be just the nest and nest site, or may include the entire home range.

Territoriality has many functions, such as the formation and maintenance of pair relations, the reduction of interference with reproductive activities, and the prevention of waste of energy through direct aggression. The territory holder often 'advertises its ownership' to avoid unnecessary contacts with other individuals. Many territorial mammals use secretions (pheromones) from scent glands to mark strategic sites within the territory, while perching birds often sing from a conspicuous place.

Australian magpies breed in territories, but surplus individuals often flock outside the breeding territories. Robert Carrick, CSIRO wildlife biologist, banded some 3000 magpies near Canberra to study their territories. He discovered, among other interesting facts, that a disease which spread in the magpie population did not infect the territory holders, whereas the flocking birds, being in close contact, were adversely affected. In this case territoriality reduced the spread of epidemic among breeding populations. Territoriality as a spacing mechanism also reduces the chance of in-breeding and predation.

There are two other presumed functions of territoriality which have become controversial in recent years, particularly in the study of birds. One notion is that territory regulates the density of many populations in the most favoured areas. There is some evidence that the most suitable habitat is the first to be occupied by territorial animals and that the dominant animals usually settle in the best areas. Two Dutch zoologists, H. N. Kluyver and L. Tinbergen, found that the breeding density of great tits was much higher and much more stable in broad-leaf woodland than in pine forest. They claimed that the territorial system of this species regulated the density in the more suitable habitat of broad-leaf woodland. This conclusion was widely accepted; territories were considered to be like rubber discs — the more they are compressed, the greater becomes the resistance to further compression. Therefore, the territorial system should not set an absolute limit on the number of animals, but the population should fluctuate within a restricted range at high densities. If the conditions of the habitat remain good (e.g. abundant food supply), there should be many years in which the density would be near the limit of maximum compression.

British ornithologists, David Lack and his associates at Oxford, found that the great tit population there increased one year far above the density of the preceding year when it was considered to be near the upper limit. Moreover, the consequent decrease in size of territory was not related to the abundance of food in the area but to the increased population density, so that the young suffered greater mortality than in any other year.

The silvereye on Heron Island in the Great Barrier Reef is maintained at a very high density (at least 400 birds in about 12 hectares of available habitat). In the breeding season, territorial defence is limited normally to the nesting tree. Some birds, however, do not even defend the nest site. Quite often, the fledged young from several different nests are fed in the same tree by their respective parents, who exhibit no territorial behaviour towards one another. This is not due to general suppression of aggression, because they show much agonistic behaviour at feeding places. Perhaps this reduced aggression in territorial behaviour made it possible for them to maintain such a high density.

Another controversial aspect of territoriality is that the territory is said to ensure an adequate supply of food for the young. This is a classical interpretation of the function of territoriality put forward in 1920 by Eliot Howard, a learned amateur ornithologist in England. This theory does not apply to species whose territories do not include feeding areas.

There is no proof, however, that the territory contains approximately the amount of food needed to raise a brood. There is usually much more food available.than can be utilized by the territory holders, and occasionally not enough to support the young, as in the special case cited above (the great tit in England).

SOCIAL HIERARCHY

When individuals of a group are marked for identification, we can record the outcome of aggressive encounters among them. Such records obtained from a flock of silvereyes are shown in Fig. 106. If the relations of individuals show a dominance order in a straight line ($\alpha > \beta > \gamma > \cdots$) from the most dominant (alpha) to the most subordinate (omega), *peck rights* are established and the social order is referred to as the *peck order*.

Among domestic hens, the initial contest determines the dominant-subordinate relations between two birds, and once the dominance is established it is not normally challenged in subsequent encounters. Subordinate birds avoid a frontal view of the dominant bird by changing the angles of their heads as the dominant moves its head. They tend to keep a greater individual distance from the dominant than from other birds. When domestic fowls fight, the stronger, larger or more aggressive individual usually wins the fight, so that the peck order in a flock generally reflects the order of physical or physiological superiority of the birds. However, since agonistic behaviour varies with the motivations of the birds involved, the dominant-subordinate relations established in the initial contest may not always reflect the strength of the birds. Thus triangular relations may result in the flock (e.g. $\alpha > \beta > \gamma > \alpha$). This can also be brought about by injecting low-ranking birds with certain androgenic hormones to increase aggression. As in the case of territorial fights, prior residence gives a bird an advantage over one which fights in strange surroundings.

In pigeons and canaries, dominant-subordinate relations are not stable, and fighting between the same two birds may produce variable results. However, a dominant-subordinate relation will become apparent after the results of many agonistic encounters have been obtained.

The social organization of animals sometimes changes according to the season. Many birds are territorial in the breeding season, and flock in the non-breeding season. While in flocks they may have peck order or other

	a ♂	b ♀	c ♂	d ♀	e ♂	f ♂	g ♀	h ♀	i ♂	j ♀	Total no. of victories	%
a ♂		3	13	16	23	5	8	23	16	6	113	100
b ♀			24	23	20	6	14	37	13	12	149	98
c ♂				0	14	2	5	5	6	1	33	47
d ♀					9	0	1	1	2	5	18	28
e ♂				1		4	9	10	12	8	44	40
f ♂				3			1	5	8	6	23	57
g ♀				1				1	5	1	8	17
h ♀									4	1	5	6
i ♂				2						3	5	7
j ♀											0	0
Total no. of defeats	0	3	37	46	66	17	38	82	66	43		
%	0	2	53	72	60	43	83	94	93	100	Total encounters 398	

FIG. 106: Social hierarchy shown by a captive flock of silvereyes (AFTER KIKKAWA): the figures show the numbers of aggressive encounters in which the birds listed on the left won the fight over the birds listed on the top

social relations. Such changes are called changes of *social phases*.

LEADERSHIP

In a social group there is often a leader or a group of leaders who guide the movements of the whole group. Leadership is held by the animal who attracts others to follow, and integrates behaviour patterns of the group. The leader may be young or old, of either sex, and is not necessarily the dominant animal when a social hierarchy exists in the group.

Among invertebrates, army ants have a shifting type of leadership, as individuals occupying the apex of an advancing column change from time to time. Leadership in birds is found in the co-ordinated flights of large flocks, often formed by brolgas, geese, cormorants and ibises (Plate IIIc). Among social mammals, leadership often appears as part of the integrated organization of the society. In the case of red deer, a dominant stag has no function to play in the event of danger; an experienced hind gives signals and leads the herd to safety. In the case of rhesus monkeys, the dominant individual in the hierarchy takes leadership of the troop, and his ability to fight and lead determines the size of the territory in relation to neighbouring troops. Here we find that territoriality, social hierarchy and leadership are all integrated to perform social functions of the group.

Social organizations described here may also occur in our society. What is the significance of fences or hedges between houses when these do not physically prevent trespassers? What do school children expect from their class leader? In a *social system*, individuals recognize the roles of other individuals, and social behaviour is based on their expectation of such roles. In our society many social systems have evolved to perform specific functions. They exist between parents and children, a teacher and students, and between friends. Therefore, in everyday life we go through many social systems within the community. In a student debate, the role of a speaker is to convince others that his views are right. The function of the debate, however, may be to stimulate discussion among students. There are many other examples within our social systems in which the expected roles and the functions are different. The social organizations of animals described in this section are much more rigid in function and in time than the social systems of man.

In the other extreme, we have solitary species among higher animals. However, even the most solitary animal does not stay solitary throughout its entire life; solitary cuckoos, for example, which do not even build nests themselves, must have social contacts at mating. The brush-tailed possum in Australia is known to be solitary, but social responses between sexes occur during the mating season. The female cares for the young possum in her pouch for about 140 days and for a further period of four months until it becomes independent. *Antechinus*, a marsupial mouse of eastern Australia, is an aggressive mammal and if kept together in a small cage they often kill one another. In the field they are solitary, though their home ranges overlap. A remarkable synchronization of behaviour is shown in their mating, which is restricted to within one week. After the mating all males die in a very short period of time, so that the young will never be exposed to their aggression. As yet the social mechanisms involved in this remarkable life cycle are unknown. In the past, marsupials were considered incapable of developing social organization, but in recent years Australian mammalogists produced evidence to indicate that communication between individuals leading to a hierarchical organization or territoriality had evolved in some species. For example, the Queensland rock-wallaby *Petrogale inornata* shows a ritualized form of threat (Plate IVc) in encounters among individuals living in a colony.

In the understanding of social organizations, there is still much to be learned from the study of wild animals. Such studies may also help us understand what man is doing to himself and to his environment.

197

Selected Readings on Behaviour

Introductory Books

Broadhurst, P. L. 1963. *The Science of Animal Behaviour.* (Pelican Original).

Carthy, J. D. 1965. *The Behaviour of Arthropods.* Edinburgh.

Dethier, V. G. & Stellar, E. 1970. *Animal Behaviour, its Evolutionary and Neurological Basis.* 3rd. ed. Englewood Cliffs.

Manning, A. 1967. *Introduction to Animal Behaviour.* London.

Tinbergen, N. 1965. *Animal Behavior.* Life Nature Library. New York.

Van der Kloot, W. G. 1968. *Behavior.* New York.

Interesting Readings

Ardrey, R. 1966. *The Territorial Imperative.* New York.

Bastock, M. 1967. *Courtship: A Zoological Study.* London.

Cousteau, J. Y. 1953. *The Silent World.* London.

Griffin, D. R. 1958. *Listening in the Dark.* New Haven.

Hasler, D. 1966. *Underwater Guideposts: Homing of Salmon.* Madison.

Hediger, H. 1950. *Wild Animals in Captivity.* London; 1955. *Studies of the Psychology and Behaviour of Captive Animals in Zoos and Circuses.* London.

Lack, D. 1943. *The Life of the Robin.* London.

Lorenz, K. 1952. *King Solomon's Ring.* London; 1954. *Man Meets Dog.* London; 1966 (1963). *On Aggression.* London.

Tinbergen, N. 1953. *Herring Gull's World.* London; 1953. *Social Behaviour in Animals.* London; 1958. *Curious Naturalists.* London.

Wickler, W. 1968. *Mimicry in Plants and Animals.* London.

Collections of Articles and Simple Experimental Work

Hainsworth, M. D. 1967. *Experiments in Animal Behaviour.* London.

McGill, T. E. (ed.) 1965. *Readings in Animal Behavior.* New York.

Readings from *Scientific American.* 1967. *Psychobiology: The Biological Bases of Behavior.* San Francisco.

Stokes, A. W. (ed.) 1968. *Animal Behavior in Laboratory and Field.* San Francisco.

Advanced Texbooks

Eibl-Eibesfeldt, I. 1967. *Grundriss der vergleichenden Verhaltensforschung.* München.

Hinde, R. A. 1970. *Animal Behavior. A Synthesis of Ethology and Comparative Psychology.* 2nd. ed. New York.

Marler, P. & Hamilton, J. P. 1966. *The Mechanisms of Animal Behavior.* New York.

Thorpe, W. H. 1963. *Learning and Instinct in Animals.* 2nd. ed. London.

Zoology Textbooks
Barnes, R. D. 1963. *Invertebrate Zoology*. Philadelphia.
Parker, T. J. & Haswell, W. A. 1964. *A Text-Book of Zoology*. Vol. II. *Phylum Chordata* (revised by Marshall, A. J.). 7th. ed. with corrections. London.

General Behaviour Books
Bliss, E. L. (ed.) 1962. *Roots of Behavior*. New York.
Carthy, J. D. 1958. *An Introduction to the Behaviour of Invertebrates*. London.
Carthy, J. D. & Ehling, F. J. (ed.) 1964. *The Natural History of Aggression*. London.
Cott, H. B. 1965. *Adaptive Colouration in Animals*. 3rd. imp. London.
Eibl-Eibesfeldt, I. 1970. *Ethology: The Biology of Behavior*. Translated by E. Klinhammer. New York.
Ellis, P. E. (ed.) 1965. *Social Organization of Animal Communities*. London.
Etkin, W. (ed.) 1964. *Social Behavior and Organization among Vertebrates*. Chicago.
Fraenkel, S. G. & Gunn, D. L. 1961 (1940). *The Orientation of Animals*. Oxford.
Grasse, P.-P. (ed.) 1956. *L'instinct dans le comportement des animaux et de l'homme*. Paris.
Hirsch, J. (ed.) 1967. *Behavior: Genetic Analysis*. New York.
Klopfer, P. H. & Hailman, J. P. 1967. *An Introduction to Animal Behavior*. Englewood Cliffs.
Lanyon, W. E. & Tavolga, W. N. (ed.) 1960. *Animal Sounds and Communication*. Washington D.C.
Lorenz, K. 1965. *Evolution and Modification of Behavior*. Chicago.
Parsons, P. A. 1967. *The Genetic Analysis of Behaviour*. London.
Roe, A. & Simpson, G. G. (ed.) 1958. *Behavior and Evolution*. New Haven.
Schiller, C. H. (ed.) 1957. *Instinctive Behavior*. New York.
Sebeck, T. A. (ed.) 1968. *Animal Communication*. Bloomington.
Thorpe, W. H. & Zangwill, O. L. 1961. *Current Problems in Animal Behaviour*. Cambridge.
Tinbergen, N. 1969 (1951). *The Study of Instinct*. Oxford.
Verplanck, W. S. *A Glossary of Some Terms used in the Objective Science of Behavior*. Washington D.C.
Wynne-Edwards, V. C. 1962. *Animal Dispersion in Relation to Social Behaviour*. Edinburgh.

Treatment of Particular Animal Groups
Altmann, S. A. (ed.) 1967. *Social Communication among Primates*. Chicago.
Armstrong, E. A. 1965. *Bird Display and Behavior*. New York.
Barends, G. P. & Barends-von Roon, J. M. 1967. *An Introduction to the Study of the Ethology of Cichlid Fish*. Leiden.
Bertrand, M. 1969. *The Behavioral Repertoire of the Stumptail Macaque*. Basel.
Carpenter, C. R. (ed.) 1969. Proceedings of Second International Congress of Primatology. Vol. I. *Behavior*. Basel.
Collias, N. A. & Collias, E. C. 1965. *Evolution of Nest-building in the Weaverbirds (Ploceidae)*. Cambridge.
Darling, F. 1937. *A Herd of Red Deer*. Oxford.

DeVore, I. (ed.) 1965. *Primate Behavior: Field Studies of Monkeys and Apes.* New York.

Eastwood, E. 1967. *Radar Ornithology.* London.

Etkin, W. 1967. *Social Behavior from Fish to Man.* Chicago.

Ewer, R. F. 1968. *Ethology of Mammals.* London.

von Frisch, K. 1968. *The Dance Language and Orientation of Bees.* London.

Hafez, E. S. E. (ed.) 1962. *The Behaviour of Domestic Animals.* London.

Hinde, R. A. (ed.) 1969. *Bird Vocalizations: Their Relation to Current Problems in Biology and Psychology.* Cambridge.

Johnsgard, P. A. 1966. *Handbook of Waterfowl Behaviour.* Oxford.

Jolly, A. 1966. *Lemur Behavior: A Madagascar Field Study.* Chicago.

Klüver, H. 1957. *Behavior Mechanisms in Monkeys.* Chicago.

Kummer, H. 1968. *The Social Organisation of Hamadryas Baboons.* Basel.

Matthews, G. V. T. 1968. *Bird Navigation.* 2nd. ed. Cambridge.

Morris, D. (ed.) 1967. *Primate Ethology.* London.

Montagu, M. F. A. (ed.) 1968. *Man and Aggression.* New York.

Norris, K. S. (ed.) 1966. *Whales, Dolphins and Porpoises.* Berkeley.

Rheingold, H. D. (ed.) 1963. *Maternal Behavior in Mammals.* New York.

Rosenblum, L. A. (ed.) 1968. *The Squirrel Monkey.* New York.

Sakagami, S. & Michener, C. D. 1962. *The Nest Architecture of the Sweat Bees (Halictinae): A Comparative Study of Behavior.* Lawrence.

Schaller, G. B. 1963. *The Mountain Gorilla, Ecology and Behavior.* Chicago.

Schrier, A. M., Harlow, H. F. & Stollnitz, F. (ed.) 1965. *Behavior of Nonhuman Primates.* 2 vols. New York.

Scott, J. P. & Fuller, J. L. 1965. *Genetics and the Social Behavior of the Dog.* Chicago.

Smith, D. D. 1965. *Mammalian Learning and Behaviour.* Cambridge.

Southwick, C. H. (ed.) 1963. *Primate Social Behavior.* Princeton.

Sudd, J. H. 1967. *An Introduction to the Behaviour of Ants.* London.

Thorpe, W. H. 1961. *Bird Song.* Cambridge.

Wells, M. J. 1962. *Brain and Behaviour in Cephalopods.* London.

Wheeler, W. M. 1928. *The Social Insects: Their Origin and Evolution.* New York.

Witt, P. N., Reed, C. F., & Peakall, D. B. 1968. *A Spider's Web, Problems in Regulating Behavior.* New York.

Zuckerman, S. 1932. *The Social Life of Monkeys and Apes.* London.

Physiological Approach

Altman, J. 1966. *Organic Foundations of Animal Behavior.* New York.

Beaument, J. W. L. (ed.) 1962. *Biological Receptor Mechanisms.* Cambridge.

Bourne, G. H. (ed.) 1968-69. *The Structure and Function of Nervous Tissue.* 3 vols. New York.

Brazier, M. A. B. (ed.) 1960. *The Central Nervous System and Behavior.* New York.

Bullock, T. H. & Horridge, G. A. 1965. *Structure and Function in the Nervous Systems of Invertebrates.* 2 vols. San Francisco.

Busnel, R. G. (ed.) 1963. *Acoustic Behaviour of Animals.* Amsterdam.

Camougis, G. 1970. *Nerve, Muscle, and Electricity.* New York.

Carlson, F. D. (ed.) 1968. *Physiological and Biochemical Aspects of Nervous Integration*. Englewood Cliffs.

Cloudsley-Thompson, L. J. 1962. *Rhythmic Activity in Animal Physiology and Behaviour*. London.

Delafresnaye, F. J. (ed.) 1961. *Brain Mechanisms and Learning*. Oxford.

Danielle, J. F. & Brown, R. (ed.) 1967. *Physiological Mechanisms in Animal Behaviour*. Cambridge.

Eccles, J. C. 1953. *The Neurophysiological Basis of Mind*. Oxford.

Eiduson, S., Geller, E., Yuwiler, A. M., Eiduson, B. T. 1964. *Biochemistry and Behavior*. Princeton.

Gaito, J. (ed.) 1966. *Macromolecules and Behaviour*. Amsterdam.

Gellhorn, E. 1953. *Physiological Foundations of Neurology and Psychiatry*. Minneapolis.

Glaser, E. M. 1966. *The Physiological Basis of Habituation*. London.

Granit, R. 1955. *Receptors and Sensory Perception*. New Haven.

Griffith, J. S. 1967. *A View of the Brain*. Oxford.

Hoagland, H. (ed.) 1957. *Hormones, Brain Function and Behavior*. New York.

Hodgson, E. 1968. *Neurobiology and Animal Behavior*. Foresman.

Horridge, G. A. 1968. *Interneurons: Their Origin, Action, Specificity, Growth, and Plasticity*. San Francisco.

Hughes, G. M. (ed.) 1966. *Nervous and Hormonal Mechanisms of Integration*. New York.

Ingle, D. (ed.) 1968. *The Central Nervous System and Fish Behavior*. Chicago.

Jasper, H. H., Proctor, L. D., Knighton, R. S., Nashay, W. C., & Costello, R. T. 1958. *Reticular Formation of the Brain*. London.

Lynn, R. 1966. *Attention, Arousal and the Orientation Reaction*. Oxford.

Klemm, W. R. 1969. *Animal Electroencephalography*. New York.

Martini, L. & Ganong, W. F. (ed.) 1966-67. *Neuroendocrinology*. 2 vols. New York.

McCashland, B. W. 1968. *Animal Coordinating Mechanisms*. Dubugue.

McCleary, R. A. & Moore, R. Y. 1965. *Subcortical Mechanisms of Behavior*. New York.

McKerrus, K. W. (ed.) 1968. *Functions of the Adrenal Cortex*. 2 vols. The Hague.

Ochs, S. 1965. *Elements of Neurophysiology*. New York.

Palmer, A. C. 1965. *Introduction to Animal Neurology*. Oxford.

Prasad, M. R. N. 1969. *Progress in Comparative Endocrinology*. New York.

Roeder, K. D. 1967. *Nerve Cells and Insect Behavior*. 2nd. ed. London.

Rosenblith, W. A. (ed.) 1961. *Sensory Communication*. New York.

Schmitt, F. O. (ed.) 1962. *Macromolecular Specificity and Biological Memory*. Cambridge, Mass.

Sheer, D. E. (ed.) 1961. *Electrical Stimulation of the Brain: An Interdisciplinary Survey of Neurobehavioral Integrative Systems*. Austin.

Slater, L. (ed.) 1963. *Bio-Telemetry*. New York.

Woodburne, L. S. 1967. *The Neural Basis of Behavior*. Columbus.

Wurtman, R. J., Axelrod, J. & Kelly, D. E. 1969. *The Pineal*. New York.

Young, J. Z. 1966. *The Memory System of the Brain*. Oxford.

Zarrow, M. X., Yochim, J. M., McCarthy, V. L. & Sanborn, R. C. 1964. *Experimental Endocrinology*. New York.

Psychological Approach

Barnett, S. A. 1967. *The Rat: A Study in Behaviour*. Aldine.

Beach, F. A. (ed.) 1965. *Sex and Behavior*. New York.

Birney, R. C. & Teevan, R. C. (ed.) 1961. *Instinct*. Princeton.

Bowlby, J. 1969. *Attachment and Loss*. I. *Attachment*. London.

Braun, J. R. (ed.) 1963. *Contemporary Research in Learning*. Princeton.

Brown, R., Galanter, E., Hess, E. H., & Mandler, G. (ed.) 1962. *New Directions in Psychology*. New York.

Grose, R. F. & Birney, R. C. (ed.) 1963. *Transfer of Learning*. Princeton.

Gross, C. G. & Zeigler, H. P. (ed.) 1969. *Readings in Physiological Psychology*. 3 vols. London.

Hebb, D. O. 1949. *The Organization of Behavior*. New York.

Hooker, D. 1952. *The Prenatal Origin of Behavior*. Lawrence.

Jones, M. R. (ed.) 1959. *Current Theory and Research on Motivation*. Lincoln, Nebraska.

Kuo, Z. Y. 1967. *The Dynamics of Behavior Development*. New York.

Lashley, K. S. 1929. *Brain Mechanisms and Intelligence*. Chicago.

Maier, R. A. & Maier, B. M. 1970. *Comparative Animal Behavior*. Belmont.

Maier, N. R. F. & Schneirla, T. D. 1965. *Principles of Animal Psychology*. New York.

Mowren, O. H. 1960. *Learning Theory and Behavior*. New York.

National Research Council (U.S.) 1967. *Communication Systems and Resources in Behavioral Sciences*. Washington D.C.

Riopelle, A. J. (ed.) 1967. *Animal Problem Solving*. Harmondsworth.

Skinner, B. F. 1953. *Science and Human Behavior*. New York.

Sluckin, W. 1965. *Imprinting and Early Learning*. London.

Spence, K. W. 1956. *Behavior Theory and Conditioning*. New Haven.

Stevenson, H. W., Hess, E. S. & Rheingold, H. L. 1967. *Early Behavior: Comparative and Developmental Approaches*. New York.

Thompson, R. F. 1968. *Foundations of Physiological Psychology*. London.

Thompson, T. & Schuster, C. R. 1968. *Behavioral Pharmacology*. Englewood Cliffs.

Tolman, E. C. 1958. *Behavior and Psychological Man*. Berkeley.

Weiskrantz, W. (ed.) 1968. *Analysis of Behavioral Change*. London.

Young, P. T. 1961. *Motivation and Emotion*. New York.

Periodicals

Animal Behaviour (previously *British Journal of Animal Behaviour*) 1953–

Animal Behaviour Monographs 1968–

Behavioral Science 1956–

Behaviour: An international journal of comparative ethology 1947–

Biology of Reproduction 1969–

Brain, Behavior and Evolution 1968–

Hormones and Behavior 1969–

Insectes Sociaux 1954–

Journal of Comparative and Physiological Psychology 1947–

Journal of the Experimental Analysis of Behavior 1958–

Physiology and Behavior 1966–

Zeitschrift für Tierpshychologie 1937–

Irregular Publications and Non-Behavioural Journals containing Behaviour Articles
Physiological Zoology, Ecology, Ibis, Auk, Journal of Mammalogy, Primates (Journal of Primatology), Viewpoints in Biology, Annual Review of Psychology, Quarterly Review of Biology, Brain Function, Communications in Behavioral Biology, Contributions to Sensory Physiology, Progress in Physiological Psychology, Journal of Animal Ecology.

Chapter References

CHAPTER 2

Beach, F. A. 1950. The Snark was a Boojum. *Amer. Psychol.* **5**, 115-24.

Blurton Jones, N. G. 1968. Observations and experiments on causation of threat displays of the great tit *(Parus major)*. *Anim. Beh. Monogr.* **1**, 75-158.

Erickson, C. J. and Lehrman, D. S. 1964. Effect of castration of male ring doves upon ovarian activity of females. *J. comp. Physiol. Psychol.* **58**, 164-6. See also Lehrman in Bliss, K. L. (ed.) 1962 (Selected Readings under 6), and in Beach, F. A. (ed.) 1965 (Selected Readings under 9).

Espinas, A. 1878. *Des Societés Animales.* 2nd ed. Paris.

Fabre, J. H. 1879-1907. *Souvenirs Entomologiques.* 10 vols. Paris. See Fabre, J. H. 1916. *The Hunting Wasps.* London.
    1920. *The Wonders of Instinct.* London.

Griffin, see Griffin, D. R. 1958 (Selected Readings under 2).

Kikkawa, J. 1966. Populations of Animals. In *Ecology and Evolution* (Modern Biology 9), ed. by Kawanabe, H. *et al.*, Tokyo (in Japanese).

Lashley, see Lashley, K. S. 1929 (Selected Readings under 9).

Lissmann, H. W. 1963. Electric location by fishes. *Sci. Amer.* March 1963, 50-9.

Loeb, J. 1918. *Forced Movements, Tropisms and Animal Conduct.* New York.

Lorenz, see Schiller, C. H. (ed.) 1957 (Selected Readings under 6).

Morgan, C. L. 1891. *Animal Life and Intelligence.* Boston.
    1906. *An Introduction to Comparative Psychology.* 2nd ed. New York.

Pavlov, I. P. 1927. *Conditioned Reflexes: An Investigation of the Physiological Activity of the Cerebral Cortex.* English ed. London.

Robertson, D. R. 1968. A comparative study of the reproductive behaviour of Australian gudgeons (Pisces: Eleotridae). B.Sc. Honours Thesis, University of Queensland Biology Library.

Romanes, G. J. 1884. *Mental Evolution in Animals.* London.
    1889. *Mental Evolution in Man.* New York.

Rowley, I. 1967. A fourth species of Australian corvid. *Emu* **66**, 191-210.

Spencer, H. 1890. *Principles of Psychology.* 2 vols. 3rd ed. London.

Sutherland, N. S. 1959. A test of a theory of shape discrimination in *Octopus vulgaris* Lamarck. *J. comp. Physiol. Psychol.* **52**, 135-41.

Sutherland, N. S., Mackintosh, J. and Mackintosh, N. J. 1963. The visual discrimination of reduplicated patterns by *Octopus*. *Anim. Behav.* **11**, 106-10.

Thorndike, E. L. 1911. *Animal Intelligence.* New York.

Tinbergen, see Tinbergen, N. 1969 (1951) (Selected Readings under 6).

Von Uexküll, J., see Schiller, C. H. (ed.) 1957 (Selected Readings under 6).

Watson, J. B. 1930. *Behaviorism.* New York.

Behavioural adaptation of Australian birds:

Immelmann, K. 1963. Drought adaptations in Australian desert birds. *Proc. XIII Int. Orn. Congr.*, 649-57.

Rowley, I. 1965. The life history of the superb blue wren *Malurus cyaneus. Emu* **64**, 251-97.

Rowley, I. 1968. Communal species of Australian birds. *Bonn. Zool. Beitr.* **19**, 362-8.

Behavioural adaptation of Australian frogs:

Main, A. R., Littlejohn, M. J., and Lee, A. K. 1959. Ecology of Australian frogs. In *Biogeography and Ecology in Australia,* ed. by Keast, A. *et al.*, Den Haag.

Adaptation of the red kangaroo:

Newsome, A. E. 1965. The distribution of red kangaroos, *Megaleia rufa* (Desmarest), about sources of persistent food and water in central Australia. *Aust. J. Zool.* **13**, 289-99.

Newsome, A. E., Stephens, D. R., and Shipway, A. K. 1967. Effect of a long drought on the abundance of red kangaroos in central Australia. *C.S.I.R.O. Wildl. Res.* **12**, 1-8.

Sharman, G. B. and Clark, M. J. 1967. Inhibition of ovulation by the corpus luteum in the red kangaroo, *Megaleia rufa. J. Reprod. Fert.* **14**, 129-37.

Bower-birds:

Marchall, A. J. 1954. *Bower-birds.* Oxford.

Drosophila behaviour genetics:

Reed, S. C. and Reed, E. W. 1950. Natural selection in laboratory populations of *Drosophila.* II. Competition between a white-eye gene and its wild type allele. *Evolution* **4**, 34-42. See also Hirsch, J. (ed.) 1967 (Selected Readings under 6).

Head-scratching in birds:

Brereton, J. Le Gay and Immelmann, K. 1962. Head-scratching in the Psittaciformes. *Ibis* **104**, 169-74.

CHAPTER 3

Adrian, E. D. 1964 (1928). *The Basis of Sensation.* London.

Buchsbaum, R. M. 1938. *Animals without Backbones.* Chicago.

Couteaux, R. 1960. Motor end-plate structure. In *The Structure and Function of Muscle* vol. 1, ed. by Bourne, G. H., New York.

Florey, E. 1966. *An Introduction to General and Comparative Physiology.* Philadelphia.

Fretter, V. and Graham, A. 1962. *British Prosobranch Molluscs.* London.

Gelber, B. 1964. Studies of the behaviour of *Paramecium aurelia. Anim. Behav. Suppl.* **1**, 21-9.

Giersberg, H. 1928. Über den Zusammenhang zwischen morphologischem und physiologischem Farbwechsel der Stabheuschrecke *Dixippus (Carausius) morosus. Z. vergl. Physiol.* **7**, 657-95. See Florey, E. 1966 (above).

Harden-Jones, R. F. 1956. The behaviour of minnows in relation to light intensity. *J. exp. Biol.* **33**, 271-81.

Hyman, L. H. 1951. *The Invertebrates* vol. II. *Platyhelminthes* and *Rhynchocoela.* New York.

Loewenstein, W. R. 1960. Biological transducers. *Sci. Amer.* August 1960, 98-108.

Maschwitz, U. 1964. Gefahrenalarmstoffe und Gefahrenalarmierung bei sozialen Hymenopteren. *Z. vergl. Physiol.* **47**, 596-655. See Lindauer, M. 1965 in *The Physiology of Insecta* vol. II, ed. by Rockstein, M., New York.

McCashland, see McCashland, B. W. 1968 (Selected Readings under 8).

Naitoh, Y. and Eckert, R. 1969. Ionic mechanisms controlling behavioural responses of *Paramecium* to mechanical stimulation. *Science* **164**, 963-5.

Ramsay, J. A. 1952. *Physiological Approach to the Lower Animals.* Cambridge.

Sleigh, M. A. 1962. *The Biology of Cilia and Flagella.* Oxford.

Taylor, 1920. Demonstration of the function of the neuromotor apparatus in *Euplotes* by the method of microdissection. *Univ. Calif. Pub. Zool.* **19**, 403-70. See Bullock, T. H. and Horridge, G. A. 1965 (Selected Readings under 8).

Wells, M. 1968. *Lower Animals.* London.

Wood, D. W. 1968. *Principles of Animal Physiology.* London.

Young, J. Z. 1939. Fused neurons and synaptic contacts in the giant nerve fibres of cephalopods. *Philos. Trans.* **229B**, 465-503.

Hormonal control of behaviour in the hawk moth:

Markl, H. and Lindauer, M. 1965. Physiology of insect behavior. In *The Physiology of Insecta* vol. II, ed.by Rockstein, M., New York.

Testosterone injection into rat brain:

Fisher, A. E. 1964. Chemical stimulation of the brain. *Sci. Amer.* June 1964, 60-8.

CHAPTER 4

Barnes, see Barnes, R. D. 1963 (Selected Readings under 5).

Batham, E. J. and Pantin, C. F. A. 1954. Slow contraction and its relation to spontaneous activity in the sea anemone *Metridium senile* (L.). *J. exp. Biol.* **31**, 84-103.

Batham, E. J., Pantin, C. F. A., and Robson, E. A. 1960. The nerve-net of the sea anemone *Metridium senile*; the mesenteries and the column. *Quart. J. micr. Sci.* **101**, 487-510.

Bullock and Horridge, see Bullock, T. H. and Horridge, G. A. 1965 (Selected Readings under 8).

Clark, R. B. 1964. The learning abilities of nereid polychaetes and the role of the supra-oesophageal ganglion. *Anim. Behav. Suppl.* **1**, 89-100.

Hadenfeldt, D. 1929. Das Nervensystem von *Stylochoplana maculata* und *Notoplana atomata. Z. wiss. Zool.* **133**, 586-638. See Hyman, L. H. 1951 (below).

Hanström, B. 1928. *Vergleichende Anatomie des Nervensystems der wirbellosen Tiere.* Berlin. See Bullock, T. H. and Horridge, G. A. 1965 (Selected Readings under 8, vol. I).

Horridge, G. A. 1962. Learning of leg position by the ventral nerve cord in headless insects. *Proc. Roy. Soc. B.* **157**, 33-52.

Hyman, L. H. 1951. *The Invertebrates* vol. II. *Platyhelminthes and Rhynchocoela.* New York.

Hyman, L. H. 1967. *The Invertebrates* vol. VI. *Mollusca I.* New York.

Imms, A. D. 1957. *A General Textbook of Entomology.* 9th ed. London.

Parker, G. H. 1919. *The Elementary Nervous System.* Philadelphia.

Passano, L. M. 1963. Primitive nervous systems. *Proc. Nat. Acad. Sci. Wash.* **50**, 306-13.

Reisinger, E. 1926. Untersuchung am Nervensystem der *Bothrioplana semperi* Braun. *Z. Morph. Ökol. Tiere* **5**, 119-49. See Hyman, L. H. 1951 (above).

Ross, D. M. 1964. The behaviour of sessile coelenterates in relation to some conditioning experiments. *Anim. Behav. Suppl.* **1**, 43-60.

Scharrer, E. and Scharrer, B. 1963. *Neuroendocrinology.* New York.

Sedwick, A. 1898. *A Student's Text Book of Zoology* vol. 1. London. (reference for p. 75)

Sedwick, A. 1909. *A Student's Text Book of Zoology* vol. 3. London. (reference for p. 80)

Smith, J. E. 1966. The form and functions of a nervous system. In *Physiology of Echinodermata*, ed. by Boolootain, R., New York.

Snodgrass, R. E. 1952. *A Textbook of Arthropod Anatomy.* New York.

Thorne, M. J. (unpublished observations).

Wells, M. J. 1964. Learning and movement in octopuses. *Anim. Behav. Suppl.* **1**, 115-34.

Young, J. Z. 1961. Learning and discrimination in the octopus. *Biol. Rev.* **36**, 32-96. See Bullock, T. H. and Horridge, G. A. 1965 (Selected Readings under 8, vol. II).

Learning in planaria:

McConnell, J. V. 1964. Cannibals, chemicals, and contiguity. *Anim. Behav. Suppl.* **1**, 61-8.

1966. Comparative physiology: learning in invertebrates. *Ann. Rev. Physiol.* **28**, 107-36.

CHAPTER 5

Altman, see Altman, J. 1966 (Selected Readings under 8).

De Beer, G. R. 1951. *Vertebrate Zoology.* London.

Goodrich, E. S. 1930. *Studies on the Structure and Development of Vertebrates.* London.

Parker and Haswell, see Parker, T. J. and Haswell, W. A. 1964 (Selected Readings under 5).

Smythies, J. R. 1967. Brain mechanisms and behaviour. *Brain* **90**, 697-706.

Starzl, T. E., Taylor, C. W., and Magoun, H. W. 1951. Collateral afferent excitation of reticular formation of brain stem. *J. Neurophysiol.* **14**, 479-96.

Woodburne, see Woodburne, L. S. 1967 (Selected Readings under 8).

CHAPTER 6

Banks, C. J. 1957. The behaviour of individual coccinellid larvae on plants. *Brit. J. Anim. Behav.* **5**, 12-24.

Carthy, J. D. 1966 (Selected Readings under 1). See Davenport, D., Camougis, G.

and Hickok, J. F. 1960. Analyses of the behaviour of commensals in host-factor: 1. *Anim. Behav.* **8**, 99-104.

Cornetz, V. 1911. La conservation de l'orientation chez la fourmi. *Rev. Suisse Zool.* **19**, 153-73. See Fraenkel, S. G. and Gunn, D. L. 1961 (1940) (Selected Readings under 6).

Ehrenfeld, D. W. 1968. The role of vision in the sea-finding orientation of the green turtle (*Chelonia mydas*). 2. Orientation mechanism and range of spectral sensitivity. *Anim. Behav.* **16**, 281-7.

Florey, E. 1966. *An Introduction to General and Comparative Physiology*. Philadelphia.

Fraenkel and Gunn, see Fraenkel, S. G. and Gunn, D. L. 1961 (1940) (Selected Readings under 6).

Harker, J. 1964. *The Physiology of Diurnal Rhythms.* Cambridge.

Herter, K. 1927. Reizphysiologische Untersuchungen an der Karpfenlaus, *Argulus foliaceus. Z. vergl. Physiol.* **5**, 283-370. See Fraenkel, S. G. and Gunn, D. L. 1961 (1940) (Selected Readings under 6).

Kikkawa, J. 1961. Social behaviour of the white-eye *Zosterops lateralis* in winter flocks. *Ibis* **103a**, 428-42.

Lindauer, M. and Martin, H. 1963. *Naturwissenschaften* **50**, 509. See Markl, H. and Lindauer, M. 1965 in *The Physiology of Insecta* vol. II, ed. by Rockstein, M., New York.

Marler and Hamilton, see Marler, P. and Hamilton, J. P. 1966 (Selected Readings under 4).

Minnich, D. E. 1919. The photic reactions of the honey-bee, *Apis mellifera. J. exp.*

Zoöl. **52**, 293-314. See Fraenkel, S. G. and Gunn, D. L. 1961 (1940) (Selected Readings under 6).

Müller, A. 1925. Über Lichtreaktionen von Landasseln. *Z. vergl. Physiol.* **3**, 113-44. See Fraenkel, S. G. and Gunn, D. L. 1961 (1940) (Selected Readings under 6).

Ohba, S. 1954. Analysis of activity rhythm in the marine gastropod *Nassarius festivus* inhabiting the tide pool. 1. Nocturnal activity and its artificial control by light. *Biol. J. Okayama Univ.* **1**, 209-16.

Sauer, E. G. F. and Sauer, E. M. 1960. Star navigation of nocturnal migrating birds. *Cold Spr. Harb. Symp. Quant. Biol.* **25**, 463-73.

Tinbergen, see Tinbergen, N. 1969 (1951) (Selected Readings under 6).

Von Buddenbrock, W. 1922. Mechanismus der phototropen Bewegunen. *Wiss. Meersuntersuch N.F. Abt. Helgoland* **15**, 1-19. See Fraenkel, S. G. and Gunn, D. L. 1961 (1940) (Selected Readings under 6).

Von Frisch, see Lindauer, M. 1965 in *The Physiology of Insecta* vol. II, ed. by Rockstein, M., New York.

Wallraff, H. G. 1960. Does celestial navigation exist in animals? *Cold Spr. Harb. Symp. Quant. Biol.* **25**, 451-63.

Wells, M. 1968. *Lower Animals.* London.

Wigglesworth, V. B. 1941. The sensory physiology of the human louse *Pediculus humanus corporis* De Beer (Anoplura). *Parasitology* **33**, 67-109.

Wolf, E. 1927. Heimkehrvermögen der Bienen. II. *Z. vergl. Physiol.* **6**, 221-54. See Fraenkel, S. G. and Gunn, D. L. 1961 (1940) (Selected Readings under 6).

Wood, D. W. 1968. *Principles of Animal Physiology.* London.

Biological clocks:
Cloudsley-Thompson, L. J. 1962 (Selected Readings under 8).
Marler, P. and Hamilton, J. P. 1966 (Selected Readings under 4).
Chain reflex in crab feeding:
Wiersma, C. A. G. 1961. The neuromuscular system. In *The Physiology of Crustacea* vol. 2, ed. by Waterman, T. H., New York.
Corn storage by muskrats:
Errington, P. L. 1963. *Muskrat Populations.* Ames.
Disinhibition hypothesis:
Van Iersel, J. J. A. and Bol, A. C. A. 1958. Preening of two tern species: A study on displacement activities. *Behaviour* 13, 1-88.
Sevenster, P. 1961. A causal analysis of a displacement activity (fanning in *Gasterosteus aculeatus* L.). *Behaviour Suppl.* 9, 1-170.
Drift migration of Australian birds:
Kikkawa, J. (MS.) Birds of the Great Barrier Reef.
1969. Characteristics of bird distribution on Great Barrier Reef islands—a basis for conservation. *Qld. Littoral Society Newsletter* no. 30, 1-8 (mimeo.)
Ethogram of monkey:
Altmann, S. A. 1965. Sociobiology of rhesus monkeys. II. Stochastics of social communication. *J. theoret. Biol.* 8, 490-522.
Feeding cycle in Metridium:
Batham, E. J. and Pantin, C. F. A. 1950. Phases of activity in the sea-anemone, *Metridium senile* (L.), and their relation to external stimuli. *J. exp. Biol.* 27, 377-99.
Feeding in the lugworm Arenicola:
Wells, G. P. 1950. Rhythms of spontaneous activity in polychaetes. *Symp. Soc. exp. Biol.* 4, 127-42.
Klinotaxis in coccinellids:
See Banks, C. J. 1957 (above).
Locomotion control in earthworms:
See Wood, D. W. 1968. *Principles of Animal Physiology.* London.
Migration of mutton-birds:
Hitchcock, W. B. 1966. Tenth annual report of the Australian bird-banding scheme, July 1963-June 1964. *C.S.I.R.O. Division of Wildlife Research Technical Paper* 11, 36.
Orientation by green turtles:
Carr, A. 1967. Adaptive aspects of the scheduled travel of *Chelonia*. In *Animal Orientation and Navigation*, ed. by Storm, R. M., Corvallis.
Orientation by salmon:
Hasler, A. D. 1967. Underwater guideposts for migrating fishes. In *Animal Orientation and Navigation*, ed. by Storm, R. M., Corvallis.
Orthokinesis in woodlice:
Gunn, D. L. 1937. The humidity reactions of the woodlouse, *Porcellio scaber. J. exp. Biol.* 14, 178-86.
Potato-washing habits of Japanese monkeys:
Kawai, M. 1963. On the newly-acquired behaviors of the natural troop of Japanese

monkeys on Koshima Island. *Primates* **4**, 113-5.

Reversal of taxis sign in the beetle Blastophagus pinniperda:
Perttunen, V. 1958. The reversal of positive phototaxis by low temperatures in *Blastophagus pinniperda* (Col., Scolytidae). *Ann. Ent. Fenn.* **24**, 12-8.

Ritualization:
Cullen, J. M. 1966. Ritualization of animal activities in relation to phylogeny, speciation and ecology. *Phil. Trans. Roy. Soc. Lond.* B **251**, 363-74.

Tinbergen, N. 1952. Derived activities: their causation, biological significance, origin and emancipation during evolution. *Quart. Rev. Biol.* **27**, 1-32.

Statocysts and orientation in shrimps:
See Carthy, J. D. 1965 (Selected Readings under 1).

Trail marking by the fire ant Solenopsis:
Wilson, E. O. 1963. Pheromones. *Sci. Amer.* May 1963, 100-14.

CHAPTER 7
Di Cara, L. V. 1970. Learning in the autonomic nervous system. *Sci. Amer.* January 1970, 30-9.

Fantz, R. L. 1958. Pattern vision in young infants. *Psychol. Record* **8**, 43-7. See also Fantz, R. L. in *Psychobiology* 1967 (Selected Readings under 3).

Harlow, see Harlow, H. F. and Harlow, M. K. in Schrier, A. M. *et al.* (ed.) 1965 (Selected Readings under 7).

Heinroth, O. 1911. Beiträge zur Biologie, namentlich Ethologie und Psychologie der Anatiden. *Verh. 5 int. orn. Kong.*, 589-702.

Miller, N. E. and Di Cara, L. 1967. Instrumental learning of heart rate changes in curarized rats. *J. comp. Physiol. Psychol.* **63**, 12-9.

Ohgushi, R. 1955. Ethological studies on the intertidal limpets. I. Analytical studies on the homing of two species of limpets. *Jap. J. Ecol.* **5**, 31-5.

Peterson, L. R. 1966. Short-term memory. *Sci. Amer.* July 1966, 90-5.

Pribram, K. H. 1969. The neurophysiology of remembering. *Sci. Amer.* January 1969, 73-86.

Schleidt, W. M. 1961. Reaktionen von Truthühnern auf fliegende Raubvögel und Versuche zur Analyse ihrer AAM's. *Zeits. f. Tierpsychol.* **18**, 534-60.

Skinner, B. F. 1938. *The Behavior of Organisms: An Experimental Analysis.* New York.

Skinner, B. F. 1948. 'Superstition' in the pigeon. *J. exp. Psych.* **38**, 168-72.

Escape reaction of cockroach:
See Roeder, K. D. 1967 (Selected Readings under 8).

Habituation in Galeolaria:
Thorne, M. J. (unpublished observations)

Homing in Acanthozostera gemmata:
Thorne, M. J. 1968. Studies on homing in the chiton *Acanthozostera gemmata*. *Aust. J. Mar. Freshwat. Res.* **19**, 151-60.

Learning ability of birds:
Stettner, L. J. and Matyniak, K. A. 1968. The brain of birds. *Sci. Amer.* June 1968, 64-76.

CHAPTER 8

Altmann, S. A. 1968. Sociobiology of rhesus monkeys. III: The basic communication network. *Behaviour* **32**, 17-32.

Carrick, R. 1963. Ecological significance of territory in the Australian magpie *Gymnorhina tibicen. Proc. XIII int. orn. Congr.*, 740-53.

Dethier, V. G. 1957. Communication by insects: physiology of dancing. *Science* **125**, 331-6.

Dwyer, P. D. 1963. The breeding biology of *Miniopterus schreibersi blepotis* (Temminck) (Chiroptera) in north-eastern New South Wales. *Aust. J. Zool.* **11**, 219-40.

Esch, H. 1967. The evolution of the bee language. *Sci. Amer.* April 1967, 96-104.

Gordon, G. 1966. A study of movements of the short-nosed bandicoot, *Isoodon macrourus* (Gould). B.Sc. Honours Thesis, University of Queensland Biology Library.

Howard, E. 1948 (1920). *Territory in Bird Life*. London.

Kikkawa, J. 1968. Social hierarchy in winter flocks of the grey-breasted silvereye *Zosterops lateralis* Latham. *Jap. J. Ecol.* **18**, 235-46.

Kluyver, H. N. and Tinbergen, L. 1953. Territory and the regulation of density in titmice. *Arch. Néerl. Zool.* **10**, 265-80.

Lack, D. 1964. A long-term study of the great tit (*Parus major*). *J. Anim. Ecol.* **33** (Suppl.), 159-73. See also Perrins, C. M. 1965. Population fluctuations and clutch-size in the great tit, *Parus major* L. *J. Anim. Ecol.* **34**, 601-47.

Norman, J. R. 1931. *A History of Fishes*. London.

Oliver, W. R. B. 1955. *New Zealand Birds*. 2nd ed. Wellington.

Salmon, M. 1965. Waving display and sound production in the courtship behavior of *Uca pugilator;* with comparison to *U. minax* and *U. pugnax. Zoologica* **50**, 123-50.

Stimson, J. 1970. Territorial behavior of the owl limpet, *Lottia gigantea. Ecology* **51**, 113-8.

Von Frisch, see Von Frisch 1968 (Selected Readings under 7); see also Lindauer, M. in *The Physiology of the Insecta* vol. II, ed. by Rockstein, M., New York.

Wenner, A. M. and Johnson, D. L. 1967. Honeybees: do they use direction and distance information provided by their dancers? *Science* **158**, 1076-7.

Wheeler, see Wheeler, W. M. 1928 (Selected Readings under 7).

Communication among Australian finches:
Immelmann, K. 1965. *Australian Finches*. Sydney.
Dominant-subordinate relations of canaries:
Shoemaker, H. H. 1939. Social hierarchy in flocks of the canary. *Auk* **56**, 381-405.
Tsuneki, K. 1960. Social organization in flocks of canaries. *Jap. J. Ecol.* **10**, 177-89.
Dominant-subordinate relations of pigeons:
Masure, R. H. and Allee, W. C. 1934. The social order in flocks of the common chicken and the pigeon. *Auk* **51**, 306-27.
Ritchy, F. 1951. Dominance-subordination and territorial relationships in the common pigeon. *Physiol. Zool.* **24**, 167-76.

Head angles in social response of hens:
McBride, G., James, J. W. and Shoffner, R. N. 1963. Social forces determining spacing and head orientation in a flock of domestic hens. *Nature, Lond.* **197**, 1272-3.

Interspecific aggression among honeyeaters:
Immelmann, K. 1961. Beiträge zur Biologie und Ethologie australischer Honigfresser (Meliphagidae). *J. Orn.* **102**, 164-207.

Leadership in army ants:
Schneirla, T. C. 1933. Studies on army ants in Panama. *J. comp. Psychol.* **15**, 267-301.

Leadership in red deer:
Darling, F. 1937 (Selected Readings under 7).

Leadership in rhesus monkeys:
Carpenter, C. R. 1942. Sexual behavior of free-ranging rhesus monkeys *(Macaca mulatta). J. comp. Psychol.* **33**, 113-62.

Mixed species flocks of birds:
McClure, H. E. 1967. The composition of mixed species flocks in lowland and sub-montane forests of Malaya. *Wilson Bull.* **79**, 131-54.

Morse, D. H. 1970. Ecological aspects of some mixed-species foraging flocks of birds. *Ecol. Monogr.* **40**, 119-68.

Moynihan, M. 1962. The organization and probable evolution of some mixed-species flocks of neotropical birds. *Smithson. misc. Coll.* **143**(7), 1-140.

Social behaviour of marsupials:
(Antechinus) Wood, D. H. 1970. An ecological study of *Antechinus stuartii* (Marsupialia) in a south-east Queensland rain forest. *Aust. J. Zool.* **18**, 185-207.

(Brush-tailed possum) Dunnet, G. M. 1964. A field study of local populations of the brush-tailed possum *Trichosurua vulpecula* in eastern Australia. *Proc. zool. Soc. Lond.* **142**, 665-95.

(Queensland rock-wallaby) Dwyer, P. D. (Department of Zoology, University of Queensland) per. comm.

(other species) Russell, E. M. 1969. Summer and winter observations on the behaviour of the euro *Macropus robustus* (Gould). *Aust. J. Zool.* **17**, 655-64.

Schultze-Westrum, T. G. 1969. Social communication by chemical signals in flying phalangers *(Petaurus brevicepa papuanua),* in *Olfaction and Taste,* ed. by Pfaffmann, C., New York.

Social hierarchy in domestic hens:
Guhl, A. M. in Hafez, E. S. E. (ed.) 1962 (Selected Readings under 7).

Social organizations of monkeys:
Altmann, S. A. (ed.) 1967 (Selected Readings under 7).

Morris, D. (ed.) 1967 (Selected Readings under 7).

Southwick, C. H. (ed.) 1963 (Selected Readings under 7).

Symbiotic associations:
Nutman, P. S. and Mosse, B. (ed.) 1963. *Symbiotic Associations.* Cambridge.

Territories of silvereyes on Heron Island:
Wilson, J. M. 1970. Selection pressures acting on a population of *Zosterops lateralis chlorocephala* (Campbell and White) on Heron Is. during the breeding season. B.Sc. Honours Thesis, University of Queensland Biology Library.

Index

(Italic page numbers refer to diagrams)

213